Seventh International Conference on Advances in Power Electronics and Instrumentation Engineering (PEIE 2016)

Bengaluru, India
12 March 2016

Editors:

Mathivanan B
Ghanshyam Singh

ISBN: 978-1-5108-6066-7

Printed from e-media with permission by:

Curran Associates, Inc.
57 Morehouse Lane
Red Hook, NY 12571

Some format issues inherent in the e-media version may also appear in this print version.

Copyright© (2016) by Grenze Scientific Society
All rights reserved.

Printed by Curran Associates, Inc. (2018)

For permission requests, please contact Grenze Scientific Society
at the address below.

Grenze Scientific Society
Jyothi Nagar-48/1
Kesavadasapuram, Trivandrum - 695004

Phone: 0471-2360236

info@thegrenze.com

Additional copies of this publication are available from:

Curran Associates, Inc.
57 Morehouse Lane
Red Hook, NY 12571 USA
Phone: 845-758-0400
Fax: 845-758-2633
Email: curran@proceedings.com
Web: www.proceedings.com

Table of Contents

1. Design and Implementation of Cost Effective Controller for Solar PV   1-17
   Application
   Pulkit Singh and Palwalia D K

2. A New Approach of Offline Parameters Estimation for Vector Controlled   18-28
   Induction Motor Drive
   Krishna R More, Kapil P N and Hormaz Amrolia

3. A Comparative Study of Switching Strategies for Single Phase Matrix   29-38
   Converter
   Mohammadamin Yusufji Khatri and Hormaz Amrolia

Seventh International Conference on Advances in Power Electronics and Instrumentation Engineering

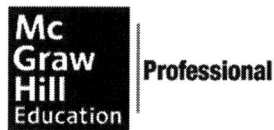

Proceedings of Seventh International Conference on Advances in Power Electronics and Instrumentation Engineering - PEIE 2016

Organized by

Preface

It is my proud privilege to welcome you all to the International Conference organized by **IDES**, held during **Mar 12, 2016 in Bengaluru, India.** This conference is jointly organized by the **IDES** and the **Association of Computer Electrical Electronics and Communication Engineers (ACEECom).** The primary objective of this conference is to promote research and developmental activities in Electrical, Electronics and Power Engineering. Another objective is to promote scientific information interchange between researchers, developers, engineers, students, and practitioners working in India and abroad.

Total 37 papers were registered for this joint International conference. Out of these 37 papers, 3 papers were registered for the Seventh International Conference on Advances in Power Electronics and Instrumentation Engineering - PEIE 2016.

I am very excited to see the research papers from various parts of the world. This proceeding brings out the various Research Papers from diverse areas of Electrical, Electronics and Power Engineering. This conference is intended to provide a common platform for Researchers, Academicians and Professionals to present their ideas and innovative practices and to explore future trends and applications in the field of Science and Engineering. This conference also provides a forum for dissemination of Experts' domain knowledge. The papers included in the proceedings are peer-reviewed scientific and practitioners' papers, reflecting the variety of Advances in Communication, Network, Electrical, Electronics, and Computing.

As a Chief Editor of this joint Conference proceeding, I would like to thank all of the presenters who made this conference so interesting and enjoyable. Special thanks should also be extended to the session chairs and the reviewers who gave of their time to evaluate the record number of submissions. To all of the members of various Committees, I owe a great debt as this conference would not have not have been possible without their constant efforts. We hope that all of you reading enjoy these selections as much as we enjoyed the conference.

Dr. B Mathivanan
Sri Ramakrishna Engineering College, India

Editorial Board

Editors-in-Chief
Dr. Mathivanan B (Sri Ramakrishna Engineering College, India)
Dr. Ghanshyam Singh (Malaviya National Institute of Technology, India)

Committees

Technical Chair
Dr. Hicham Elzabadani (American University in Dubai)
Dr. Prafulla Kumar Behera (Utkal University, India)

Technical Co-Chair
Dr. Natarajan Meghanathan (Jackson State University, USA)

Chief Editors
Dr. Ghanshyam Singh (Malaviya National Institute of Technology, India)
Dr. Mathivanan B (Sri Ramakrishna Engineering College, India)

General Chair
Dr. Janahanlal Stephen (Matha College of Technology, India)

General Co-Chair
Prof. K. U Abraham (Holykings College of Engineering, India)

Publicity Chair
Dr. Amit Manocha (Maharaja Agrasen Institute of Technology, India)

Organizing Chair
Prof. Akash Rajak (Krishna Institute of Engg. & Tech., UP, India)
Dr. Yogesh Chaba (Guru Jambeswara University, India)

Organizing Co-Chair
Dr. Theo Vande Hoak (University of Amsterdam, Holland)
Prof. Ford Lumban Gaol (University of Indonesia)

General Vice-Chair
Dr. Gylson Thomas (Jyothi Engineering College, India)

Executive Chair
Dr. Rajesh Bhatia (PEC Univeristy of Technology, India)

Program Committee Chair
Dr. Harry E. Ruda (University of Toronto, Canada)
Dr. Durga Prasad Mohapatra (NIT Rourkela, India)

Program Committee Members
Dr. Shu-Ching Chen (Florida International University, USA)
Dr. T.S.B.Sudarshan (BITS Pilani, India)
Dr. Habibollah Haro (Universiti Teknologi Malaysia)
Dr. Derek Molloy (Dublin City University, Ireland)
Prof. Jagadeesh Pujari (SDM.College of Engineering & Technology, India)
Prof. Animesh Adhikari (S P Chowgule College, India)

Dr. Anirban Mukhopadhyay (University of Kalyani, India)
Dr. Malabika Basu (Dublin Institute of Technology, Ireland)
Dr. Tahseen Al-Doori (American University in Dubai)
Dr. V. K. Bhat (SMVD University, India)
Dr. Ranjit Abraham (Armia Systems, India)
Dr. Naomie Salim (Universiti Teknologi Malaysia)
Dr. Abdullah Ibrahim (Universiti Malaysia Pahang)
Dr. Charles McCorkell (Dublin City University, Ireland)
Dr. Neeraj Nehra (SMVD University, India)
Dr. Muhammad Nubli (Universiti Malaysia Pahang)
Dr. Zhenyu Y Angz (Florida International University, USA)
Dr. Keivan Navi (Shahid Beheshti University, Tehran)
Dr. Rama Shankar Yadav (MNNIT, India)
Dr. Smriti Agrawal (MNNIT, India)
Dr. Vandana Bhattacherjee (BITS Mesra, India)
Dr. R.D. Sudhaker Samuel (S J College of Engineering, India)
Dr. Amitabha Sinha (West Bengal University of Technology, India)
Prof. SHYAM LAL (MIT MORADABAD, India)
Prof. Debasish Jena (Biju Patnaik University of Technology, India)
Prof. Srinivasa K G (M S Ramaiah Institute of Technology, India)
Dr. Bipin Bihari Jayasingh (CVR College of Engineering, India)
Dr. Seyed-Hassan Mirian-Hosseinabadi (Sharif University of Technology, Iran)
Dr. Malay K. Pakhira (Kalyani Government Engineering College, India)
Dr. Sarmistha Neogy (Jadavpur University, India)
Dr. Sreenath Niladhuri (PODICHERRY Engineering College, India)
Prof. Ananta Ojha (ICFAI University, India)
Dr. A K Sharma (YMCA Institute of Engineering, India)
Dr. Debasis Giri (IIT Kharagpur, India)
Prof. Suparna Biswas (WBUT, India)

Design and Implementation of Cost Effective Controller for Solar PV Application

Pulkit Singh* and D K Palwalia**
* Poornima Institute of Engineering & Technology Jaipur
pulkit.singh@poornima.org
** Department of Electrical Engineering, RTU Kota
dheerajpalwalia@gmail.com

Abstract: Growing pollution and depleting fossil fuel reserves have encouraged exploration and exploitation of non-conventional energy sources since oil crisis faced in late 1970. Among available renewable energy sources, photovoltaic (PV) generation is supposed to play major role in future energy scenario. High installation cost and low efficiency has been major challenges for its wide spread usage. This work focuses on reducing cost of generation from PV and enhances power tracking with low cost controller design module. Maximum power point tracking (MPPT) highly depends on atmospheric conditions and exposure of PV panel surface to solar radiation. So MPPT technique should be good enough in dynamic atmospheric conditions. Different algorithms are design in which the Perturb and Observer (P & O) and Incremental conductance (INC) are widely used MPPT techniques.

Present work focus to design the cost effective PV controller module. In the PIC technology C/C++ Coding is used to design the algorithm but the drawback of the PIC technology is the complex coding and debugging is much more difficult. Emphasizing on DsPIC30f4011, in which automatic C/C++ code for the algorithm generated in the MATLAB using simulation and can easily be dumped on the respected pins and for the completion of the coding MP-Lab IDE software is used which interface between the DsPIC chip and DsPIC hardware kit. For the compilation process complier XC16 and C30 complier is used.

Efforts have been made to design the cost effective and reliable operation of PV system using IGBT based boost converter and rectifier circuit thereby ensuring the minimum switching losses, reducing size and cost of controller. The new topology is well suited for drives and renewable energy applications.

Keywords: Solar PV, MATLAB, MPLAB, DsPIC30f4011, P&O Algorithm.

Introduction

Energy plays an important role in our daily life. With rise in dependence on electrical energy, new sources of energy need tobe explored and exploited in order to meet the energy demand. Most common sources of energy currently utilized world-wide for generating electricity includes coal (39.3%), petroleum (0.7%), natural gas (27.6%), nuclear (19.5%), hydro power (6.7%), wind (4.2%) and other renewable (2.1%) covers mainly geothermal, biomass & PV energy [1]. Fossil fuel based energy sources cause emission of carbon particle and harmful gases, causing severe environmental concerns. This has encouraged power system researchers to increase dependence on renewable energy based generation. Renewable energy is promising and inexhaustible in nature. Amongst the available natural sources, wind, solar and hydro energy are omnipresent in abundance around the globe. Wind energy is intermittent in nature, involves high initial installation cost and gets affected by the geographical condition. For terrestrial applications, solar energy has gained great attraction due to easy installation, high reliability and simplicity in design.

PV cells are used to convert solar energy into electrical energy with the help of power electronics based converters. Solar panel can easily be mounted on roof of houses, multi storage buildings and complexes. Besides these advantages, the drawback is that solar energy is not available during the night hours. So, auxiliary backup unit is necessary to assure continuous supply of electrical energy. Moreover atmospheric conditions like cloud, partial shadowed zone, dust, and snow reduces the overall efficiency of the PV cells. Kumar and Palwalia [1] discussed various MPPT algorithms used to track maximum solar power like perturb & observe (P&O), hill climbing, incremental conductance (Inc), etc [1]. In this work, a new cost effective and reliable microcontroller based P&O tracking has been implemented as MPPT. A MATLAB and MP-lab software tool has been used to obtain simulation results. The obtained results have been tested on real time environment with the help of hardware assembly and software platforms in order to design low cost and user friendly system.

Seventh International Conference on Advances in Power Electronics and Instrumentation Engineering – PEIE 2016

Aim and Objective

PV panel installation cost is quite high due to low efficiency of conversion equipment. In order to encourage its usage, cost efficient design of the system should be used. Presented the cost efficient MPPT hardware for PV module and its compatibility has been investigated on software as well as hardware platform.

The main objectives of this dissertation can be summarized as –

- Investigate working of PV module (solar cell, connections) and its application in different area by recent & relevant literature survey.
- Inspect different MPPT techniques, algorithm and converter topology depending on area of application according to their need and advancement e.g. DC- DC converter.
- Study PIC micro-controller and DsPIC, and examine pin configuration of DsPIC Kit. Mark their advantage over sensors and other methodology used for implementation.
- Analyze PV system with PIC and DsPIC Microcontroller.
- Establish software compatibility via simulation of PWM and design algorithm in MATLAB toolbox.
- Generate C language code in MATLAB and obtain hex-code with the help of MP-Lab software platform, i.e. code compilation in DsPIC.
- Incorporate obtained code in micro-controller to obtain PWM gate pulse for IGBT module connected to PV system via DsPIC kit.

Motivation

In a developing country like India, energy demand has increased exponentially in the last few decades. Majority of power is generated by conventional sources like coal and petroleum. This has increased pollution and depletion of fuel reserves. Conventional sources can deplit in short duration and takes very long time to recover. This energy generation and consumption patterns have called upon need to increase dependence on green energy sources like wind, solar, hydro, etc. High initial installation cost of renewable energy modules, energy constraints and increasing per capita energy consumption have been challenging issues for energy authorities. This need to be coped up by encouraging research programs all over the country. India has vast energy reserve in form of conventional and non-conventional sources. Energy scenario in India has been shown in table 1.

Table 1: Energy Scenarios in India

S. No	Source	% Contribution
1.	Thermal	64.75%
2.	Hydro	21.73%
3.	Nuclear	2.78%
4.	Renewable & other Sources	10.73%

Grid connected (Capacities in MW)

Table 2: Capacity of grid connected power plants

Wind Power	1234.11
Solar Power	827.22
Bio-Power	132.00
Waste to Power	12.00
Total	2311.88 MW

Off-Grid connected (Capacities in MW)

Table 3: Capacity off grid connected power plant

Waste to Energy	0.50
Biomass	10.50
Biomass Gasifiers-Rural-Industrial	0.20
	8.67
Hybrid systems	0.13
SPV Systems	46.50
Total	66.50 MW

To overcome the energy crises in power sector in last few decades, planning commission of India has set a new ministry called ministry of new and renewable energy (MNRE) with a motive to encourage use of non-conventional energy sources. MNRE has taken initiatives like Jawaharlal Nehru national solar mission (JNNSM) with an aim to install 20GW of grid connected solar system and 2 GW of off grid by year 2022. MNRE, in venture with Indian government, provides subsidy on PV module installation and provides extra benefits to people using maximum power m renewable sources.

Desgin of Photovoltaic Cell

Solar energy is the form of energy which is taken by the sun in the form of solar radiation. In other words we can define photovoltaic cells are those cells which convert solar energy direct into the electricity, using semi conductor material silicon or germanium. If we describe the background the word photo voltaic it comes from the Greek word which means "light " indicate to "photo" and "voltaic" indicate to "electrical". Efficiency of the solar cell is mainly 30 - 40%. Due to the low efficiency in the PV system different techniques are used to extract maximum power from solar panel like PIC, DSP, and FPGA & DsPIC.

Single solar cell can give only about 0.5 volt, using the solar for the terrestrial application such as home lighting, water pumping etc power generated by single solar cell is not enough. To increase the rating number of cell are connected in the series or parallel which is known as PV module. For obtaining the higher power PV panel should be connected in the array. Here the PV panels are the combination of PV modules. Series connections are responsible for increasing the voltage of the array whereas the parallel connection is responsible for increasing the current in the array.

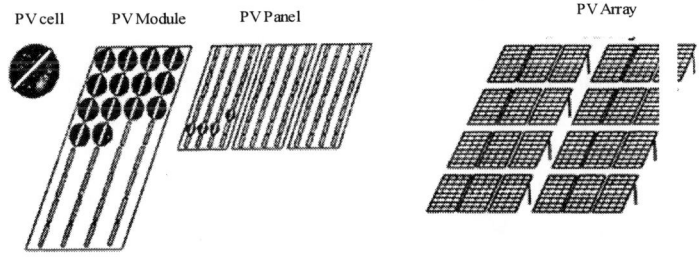

Fig 1: PV terms

The relationship between current and voltage may be determined from the diode characteristic equation that is:

$$I = I_{ph} - I_o \left(e^{qv/kt} - 1 \right) \tag{3.1}$$

$$I = I_{ph} - I_d \tag{3.2}$$

Where q is the electron charge, k is Boltzmann constant, I_{ph} is the photo current, I_o is the reverse saturation current, I_d is the diode current and T is solar cell operating temperature (K).

An ideal source can be considered as a current source where the current produce by the solar cell is proportional to the solar irradiation falling on it. But the behavior of the ideal PV cell totally changes if we consider the practical circuit. Electrical

Seventh International Conference on Advances in Power Electronics and Instrumentation Engineering – PEIE 2016

losses, optical losses are seen in the practical circuit. Modeling of the PV cell can be representation by the two type single diode model & double diode model .In the double diode model the optical losses are representation by the current sources .While the generated current I_L is directly proportional to the solar intensity. Two diode connected in the parallel representation the recombination losses. They are connected in the reverse because the recombination current flow opposite to the direction of the light generated current. Saturation current I_{S1} will flow in the diode J_{01} due the diffusion and saturation current and Saturation current I_{S2} will flow in the diode J_{o2} due to the recombination of the space charge carrier. Electrical losses (ohmic loss) which occur due to series resistance Rs and shunt resistance R_{sh} .Series resistance offered the path to the current which is flowing in the solar cell. Shunt resistance indicate the leakage path of the current in a solar cell therefore it is represented in parallel with the current sources.

I-V equation of the solar cell is given by –

$$ J = J_L - J_O (\exp \frac{q\,v}{KT} - 1) \qquad (3.3) $$

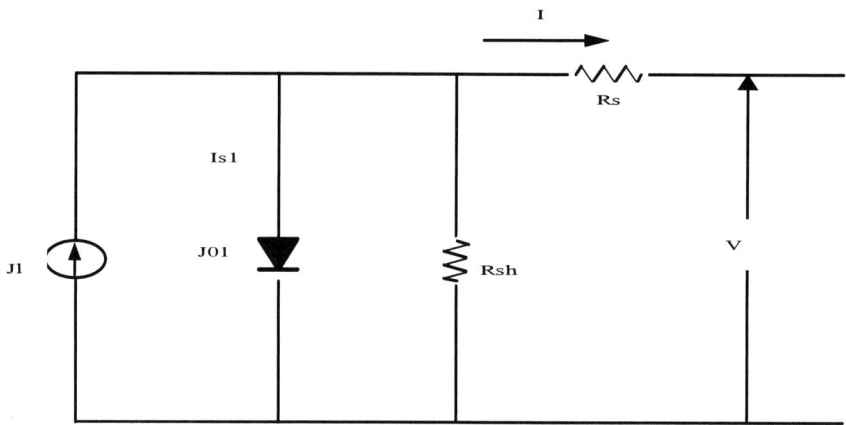

Fig 2: Single diode model

$$ J = J_l - J_{O1} \exp(\frac{q(V+IR_s)}{KT}) - J_{02} \exp(\frac{q(V+IR_s)}{KT}) - \frac{V+IR_s}{R_{sl}} \qquad (3.4) $$

Equation (3.4) represented the two diode model equation. Term J_{01} represents the recombination in base and emitter region of cell &J_{02} represent the recombination of the space charge region.

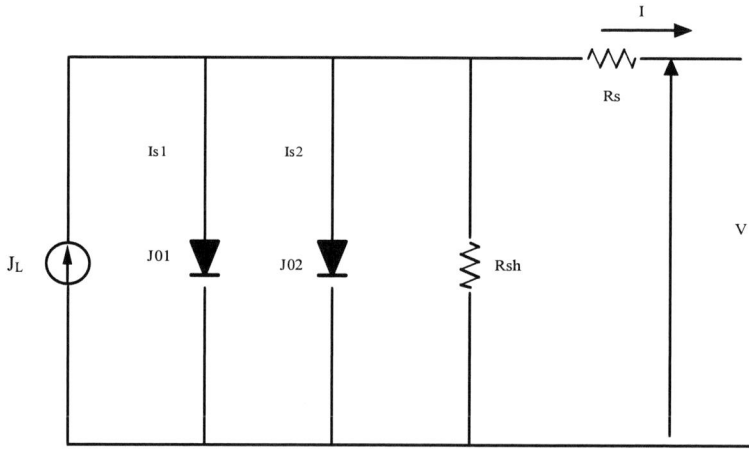

Fig 3: Double diode model

4

Design and Implementation of Cost Effective Controller for Solar PV Application

In the simple solar cell model $J_{02} = 0$. equation of the simple solar cell can be written as –

$$J = J_l - J_{O1} \exp(\frac{q(V + IR_s)}{KT}) - \frac{V + IR_s}{R_{SH}}$$

Here n is the diode ideality factor its value lie between 1 and 2 where 1 is for ideal diode and above equation represent the single diode model.

Ratings of 20w PV Module

The simulations are carried out using MATALB/SIMULINK package. The developed mathematical model of the PV array is used for the simulation studies. Various parameters of the PV array are determined and chosen. For the simulation work, we consider the solar panel model of rating 20 watt. Parameter ratings are taken from PV module datasheet

Table 4: Parameter of 20W Solar Module

Maximum Power	20 Watt
Open circuit voltage	1.2 V
Short circuit current	1.8 A
Ns	10
Np	2
Ideal factor	1.3
Band gap of semi conductor use in a cell	1.13 eV

Simulation results of 20W Solar PV Module

I-V curve of PV panel

I-V (Current-Voltage) curve originated from the equation (3.1) for particular value of the voltage .Current value we get and plot the curve this curve, shows what should be the current at a certain voltage. When the $I_{pv}=0$ we will get open circuit voltage (V_{oc}) of PV panel, when Voc=0we will get short circuit current (I_{sc}).In I-V curve represents the maximum power point corresponding voltage V_{mpp} and corresponding current I_{mpp}.

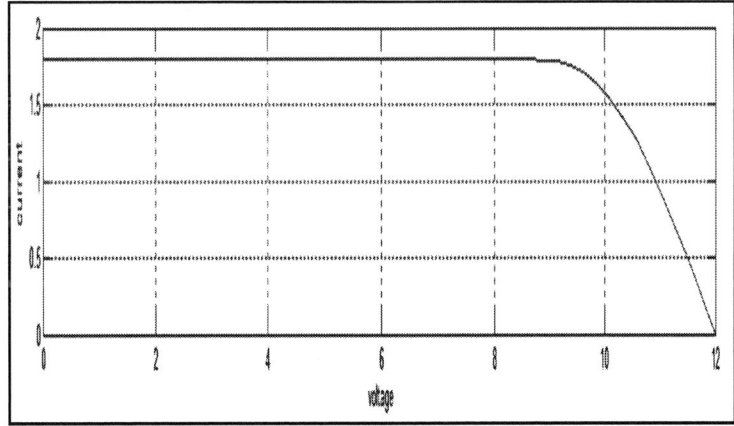

Fig 4: I-V cure at 1000W/m² irradiance

P-V curve of PV panel

Multiplication of output current and output voltage gives the output power, at particular value of Current (I_{mpp}) and voltage (V_{mpp}), will give maximum power P_{mpp}. Figure 5 shows P-V (Power-Voltage) curve of PV panel, point shows the maximum power point of the panel.

Seventh International Conference on Advances in Power Electronics and Instrumentation Engineering – PEIE 2016

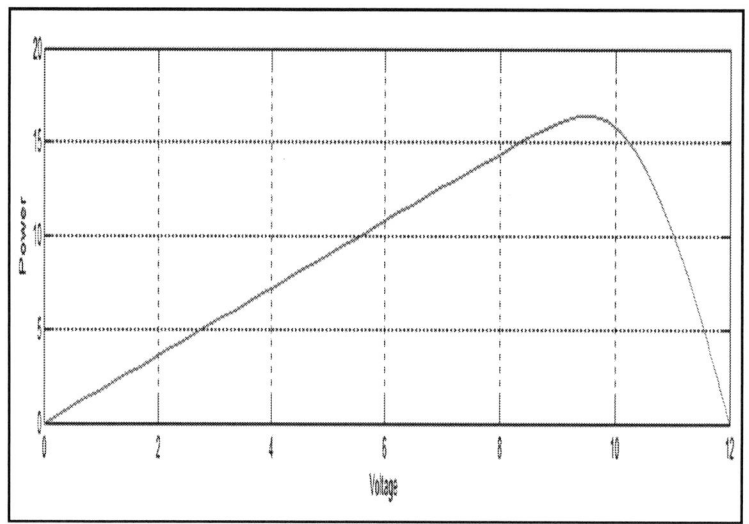

Fig 5: P-V cure at 1000W/m^2 irradiance

Varying insolation condition
Figure 6 shows the P-V curve in different insolation conditions

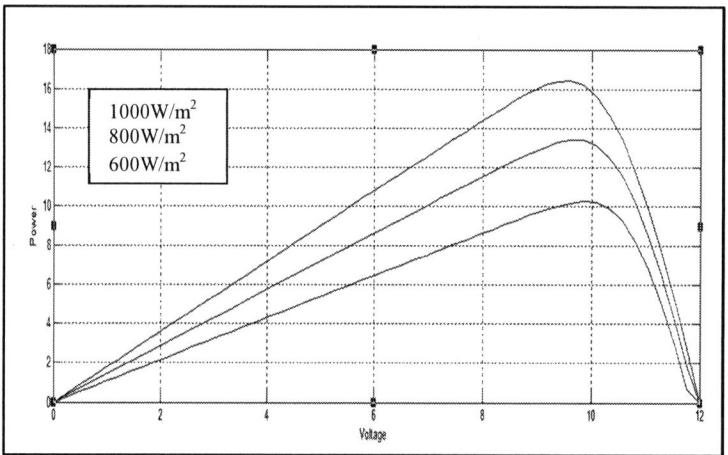

Fig 6: P-V curve at 1000W/m^2,800W/m^2,600W/m^2

Table 5: Power at different irradiance at constant temperature

Irradiance (W/m^2)	Power (watt)	Temperature (^{o}C)
1000	17.5	25
800	13	25
600	10	25

Varying temperature condition
Figure 7 & 8 shows the P-V and I-V curve in different temperature conditions, point shows peak power of each curve, as temperature increases peak power shifted downwards.

6

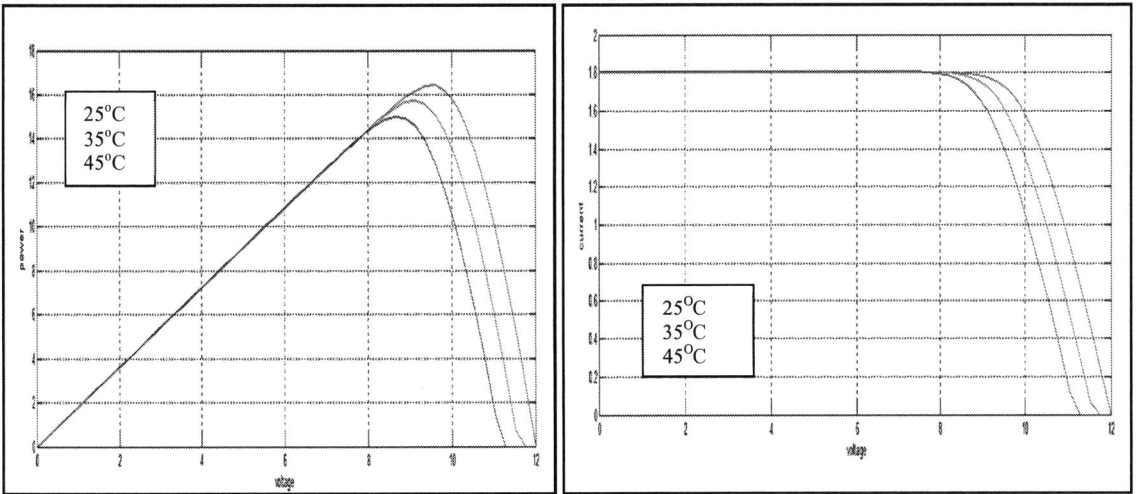

Fig 7: P-V Curve at temperature 25°C, 35°C, 45°C   Fig 8: I-V Curve at temperature 25°C, 35°C, 45°C

Maximum Power Point Tracking

Output of the solar panel varies with respect to the sun position, temperature and insulation level. Out of these conditions there are two more conditions also which affect the output power of the PV system – cloudy day and partial shadings. As the efficiency of the PV panel is low i.e. ~ 10 – 25% and if the above condition occurs then the output power of the PV panel decreases. We cannot increase the efficiency of the PV panel but we can extract the maximum power from the panel and the point at which maximum power is extracted is known as MPPT. In other words, we can define controller which tracks the maximum power point locus of the PV array/panel & it is known as MPPT.

Methods of MPPT

Algorithms of MPPT are of various types and are implemented for obtaining the maximum power. Algorithms are used in the DsPIC to implement the maximum power tracking. Different MPPT techniques are given below –
- Hill climbing/ P & O Method
- Incremental conductance Method
- Fractional open circuit voltage/ short circuit current
- Fuzzy and neural network
- dP/dV or dP/dI feedback control

Hill climbing/ P & O Method

The P & O algorithm is widely accepted algorithm due to its simplicity and easy implementation. This algorithm is also known as Hill Climbing (HC) Algorithm. The difference between the two is that only P & O algorithm will work on the PV array voltage and PV array current & hill climbing algorithm work on the concept of Duty Ratio. Although Hill climbing and P & O method are used to obtain MPP but the concept is same. In the P&O algorithm, from the I-V characterizes and conclude that the operating on the left of MPP, when there is increase in voltage then there is increase in the power while on the right hand side if voltage decreases power also decreases. Therefore with increase in power perturbation should be same to reach the desired MPP and if the power is decreased then perturbation should be positive.

Modeling of PV Module with P & O and Boost Converter

The Simulink model of the required solar cell and boost converter system is as shown. This is for the P&O method. Here the solar cell is represented by a block named 'Photovoltaic cell'. Boost converter, which consists of a 0.001H inductor and a 1F capacitor. This boost converter is used to step up the voltage to the required value. The gating signal to the boost converter is generated by comparing the signal generated by the MPPT algorithm to a repeating sequence operating at a high frequency. The load is a 10 ohm resistance P&O algorithm is applied to track MPP. Data sheet is taken from- Energy PV module 285PC8.

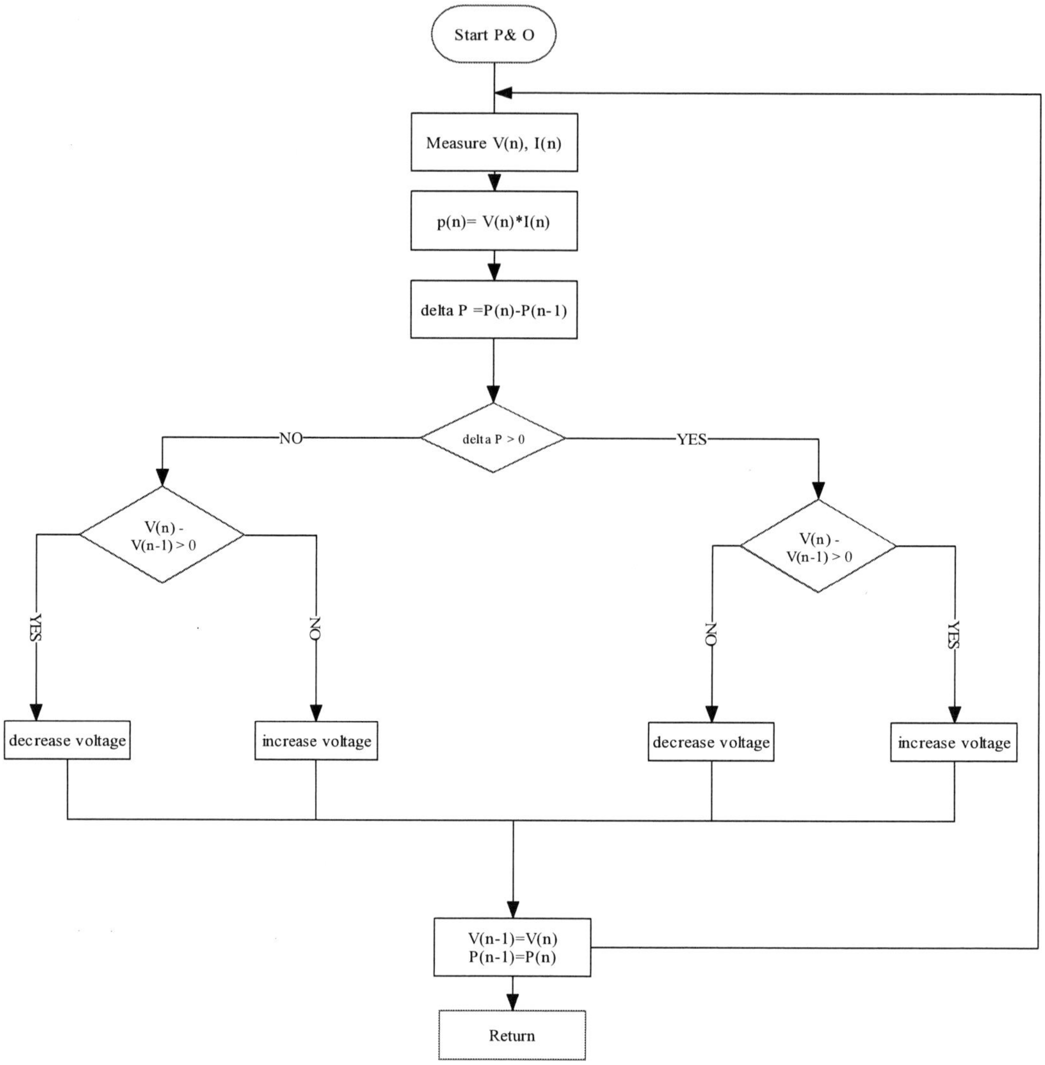

Seventh International Conference on Advances in Power Electronics and Instrumentation Engineering – PEIE 2016

Start P& O

Measure V(n), I(n)

p(n)= V(n)*I(n)

delta P =P(n)-P(n-1)

delta P > 0

NO — V(n) - V(n-1) > 0

YES — V(n) - V(n-1) > 0

YES → decrease voltage

NO → increase voltage

NO → decrease voltage

YES → increase voltage

V(n-1)=V(n)
P(n-1)=P(n)

Return

Fig 9: Flow chart of P&O Algorithm

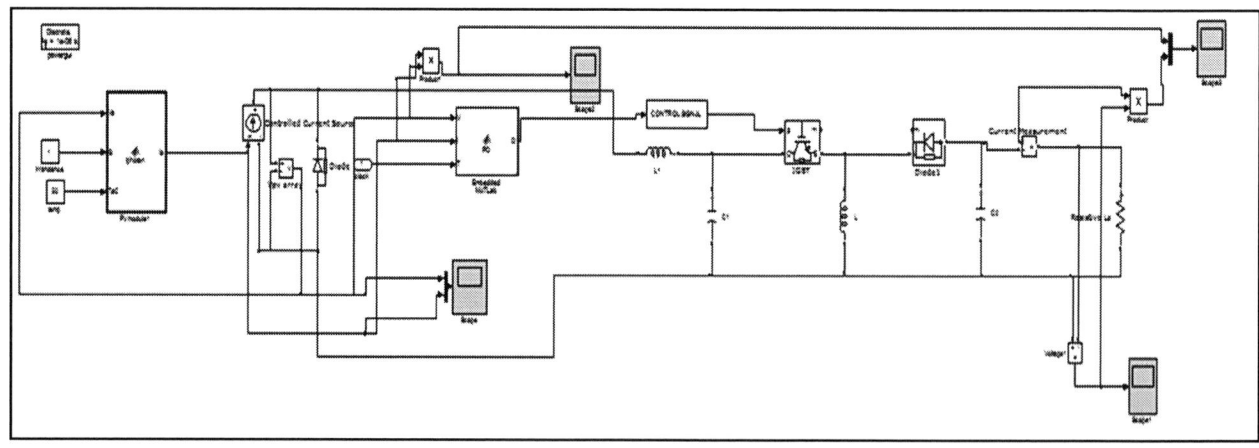

Fig 10: Simulation of PV system

Design and Implementation of Cost Effective Controller for Solar PV Application

Table 6: PV module 285PC8

Voltage at V_{max} (V_{mp})	35.9 V
Current at I_{max} (I_{mp})	7.95 A
Maximum Power	276 Watt
Open circuit voltage V_{oc}	44.5 V
Short circuit current I_{sc}	8.56 A
Module efficiency	14.37 %

Simulation result

PV module is simulated with MPPT with boost converter and output voltage, output power curve is plotted. Figure 11-16 shows input and output power.

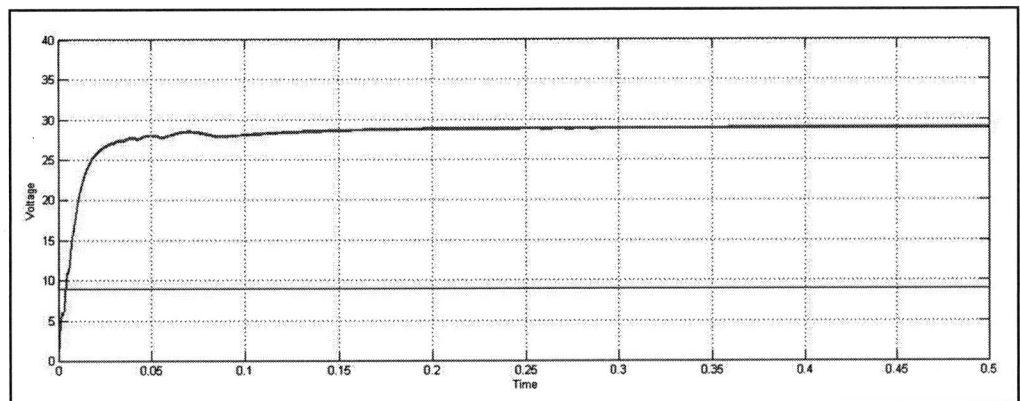

Fig 11: Input voltage 26V before boost converter

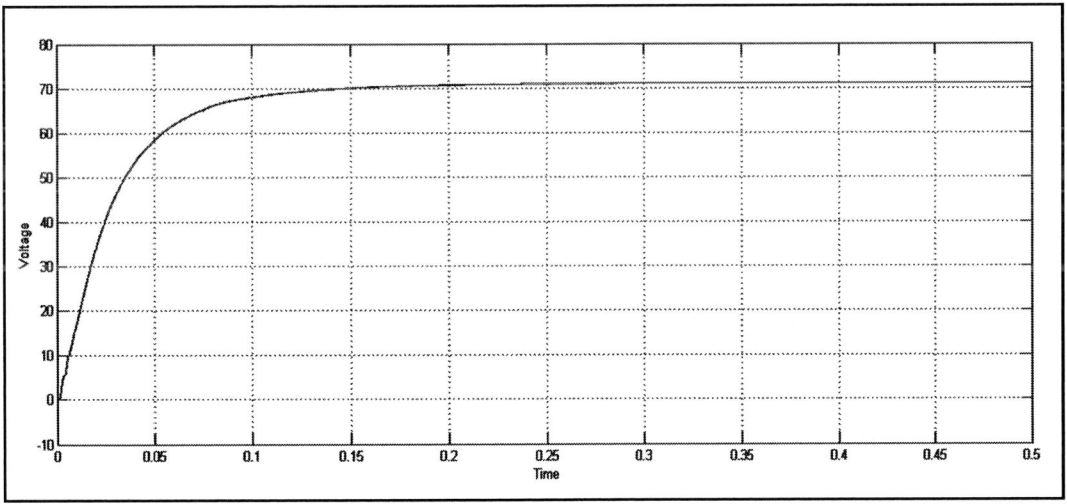

Fig 12: Output voltage 70V after boost converter

Seventh International Conference on Advances in Power Electronics and Instrumentation Engineering – PEIE 2016

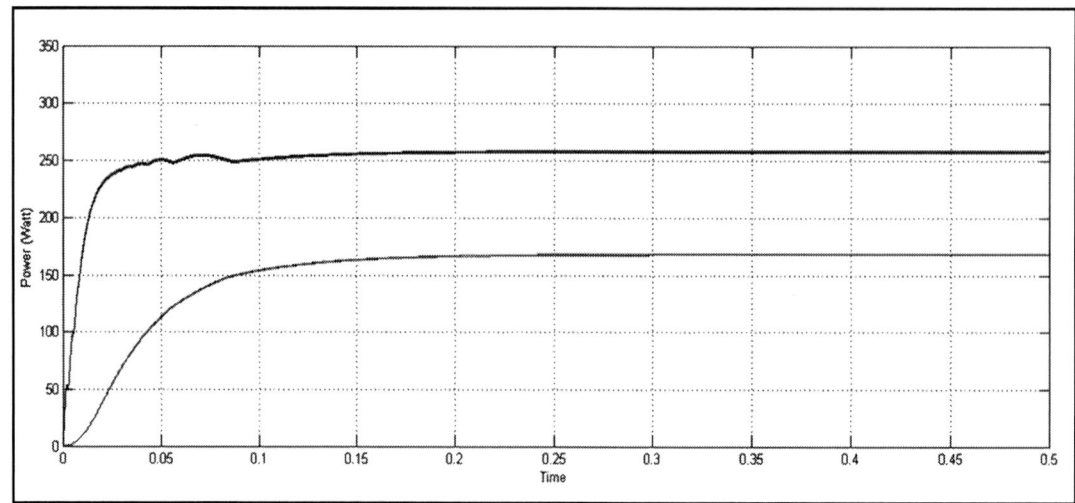

Fig 13: Output and input power at 1000W/m^2

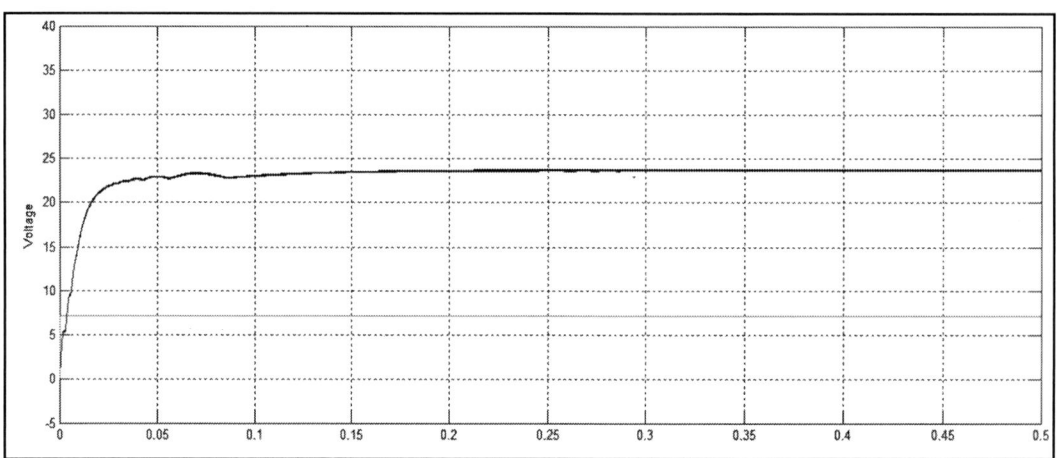

Fig 14: Input voltage 21V before boost converter

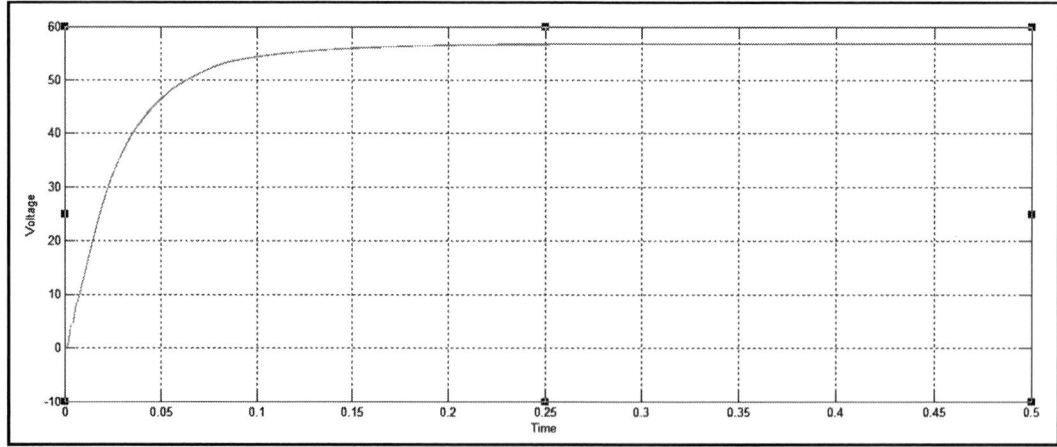

Fig 15: Output voltage 55V after boost converter

Design and Implementation of Cost Effective Controller for Solar PV Application

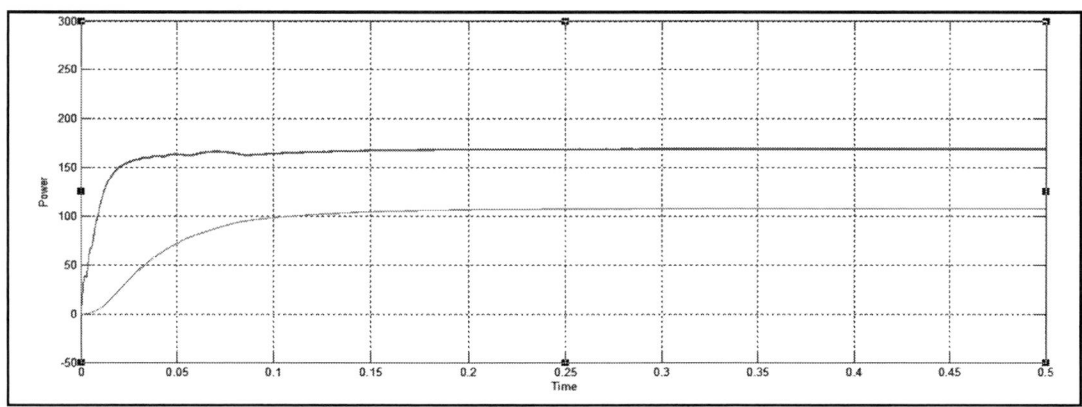

Fig 16: Input and output power at 800 W/m^2

Table 7: Simulation result

Irradiance	Input voltage	Output voltage	Input power before MPPT	Output power after MPPT
1000W/m^2	26 V	70V	150 W	250W
800W/m^2	21 V	55V	90W	150W

Harware Simultion Results

With the advantage of automatic generated code its gives another benefit of a low price which is very economical to the industry purpose. MPP is tracked using P & O Algorithm, for tracking this algorithm different medology used to track MPP and to extract the maximum power form the PV module. DsPACE, FPGA, Sensors & PIC are the different mode to track the MPP but these modules have some disadvantage which is higher cost, complex, and not user friendly system. If we consider the DsPACE and FPGA module they are they are costly and difficult to implement on the large scale. In the PIC family 18, 24 c coding is to be done in this configuration it is very difficult to find the error and error line is not shown in the system we have to find the error by checking all the line which will make very complex and lengthy. Microchip has developed a unique configuration DsPIC30F4011 which contains all the parameter to make the system low cost, easy and reliable.DsPIC30F4011 has a six PWM channel which can be utilized in the converter and automatic C/C++ code can be generated and easily dumped on the chip using MPLab. In this error finding is very simple if the simulation has some error then built will not success and error in the line can be shown in red color. This makes system easier and reliable.

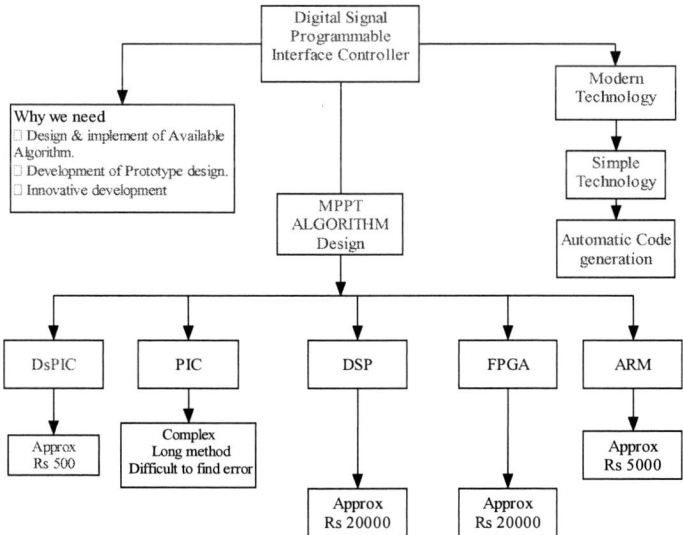

Fig 17: Flow chart showing advantage of DsPIC controller

Seventh International Conference on Advances in Power Electronics and Instrumentation Engineering – PEIE 2016

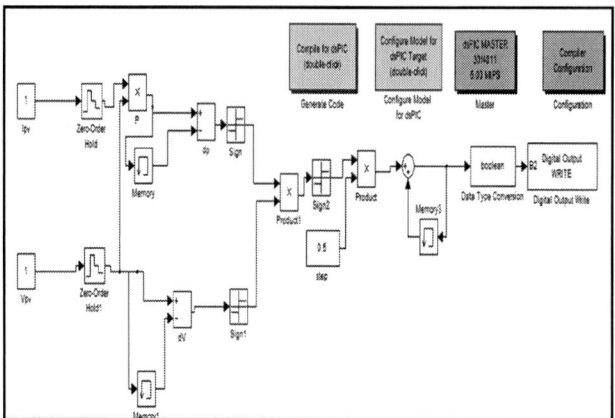

Fig 18: Simulation of P&O Algorithm using DsPIC Patch

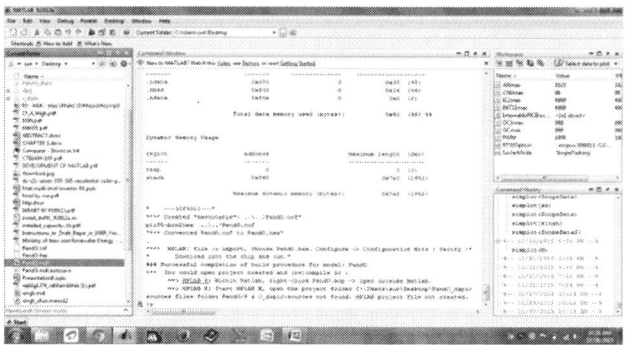

Fig 19: Successful build for P&O Algorithm

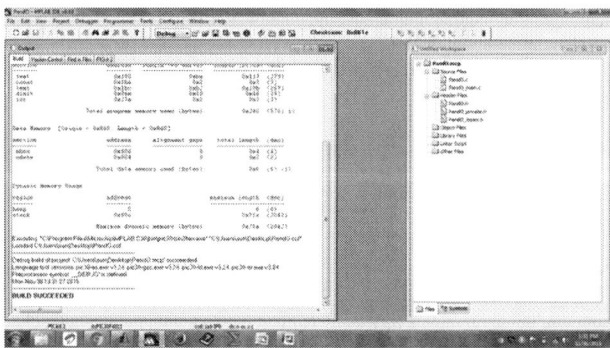

Fig 20: Successful build of code in MP Lab

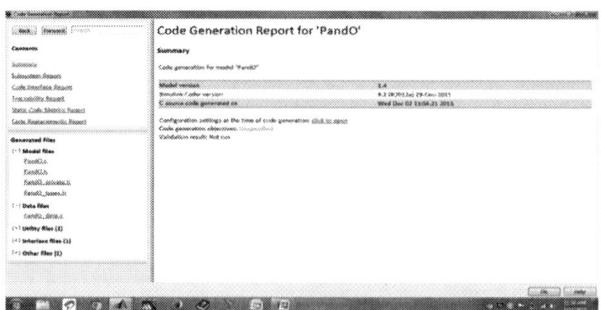

Fig 21: C Code generated report for P&O Algorithm

Design and Implementation of Cost Effective Controller for Solar PV Application

Hardware Implementation for Pv System

Hardware design of PV system is done using DsPIC30F4011. Flow chart describing the theme of the hardware.

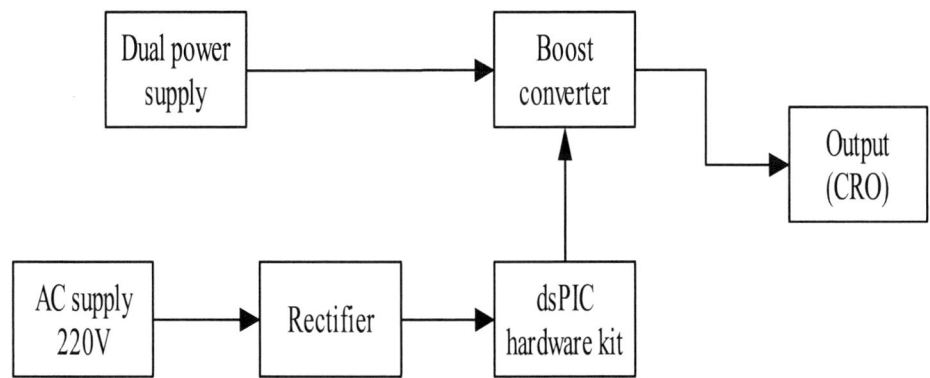

Fig 22: Block diagram showing PV system hardware

Dual power supply is used to vary the dc voltage and dual power supply act as PV module in this 9V is boost up to 24V dc using DsPIC30f4011. In the PV module as the voltage vary w.r.t to temperature and irradiances. Similarly experiment is performed by varying the voltage. Single integrated diagram is shown below-

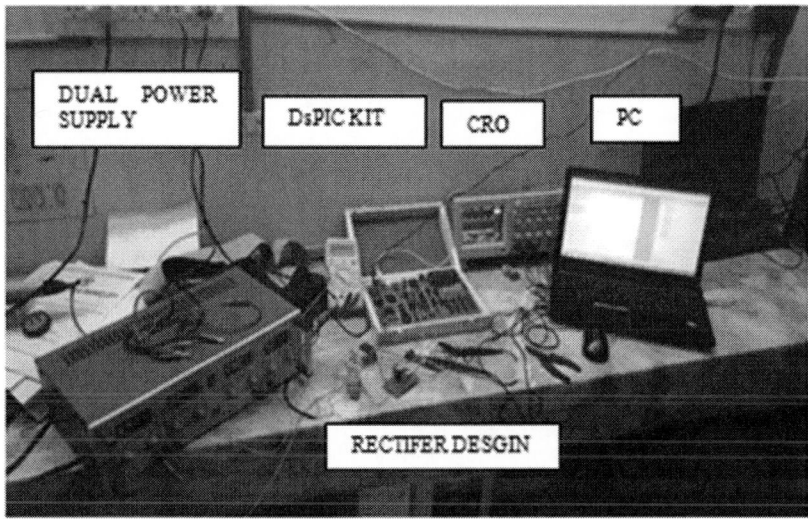

Fig 23: Hardware set up

Table 8: Components Rating

Components	Rating	Numbers
Capacitor	100uF	1
Transformer	220V-15V	1
IGBT	CT60AM-18F	1
TPL-250	3V/15V	1
IC-7825	15V	1
DsPIC30F	4011,6 Channels PWM	1
Boost converter	5V-35V	1
Resistor		2
Hardware kit		1
Diodes		4

Output waveform

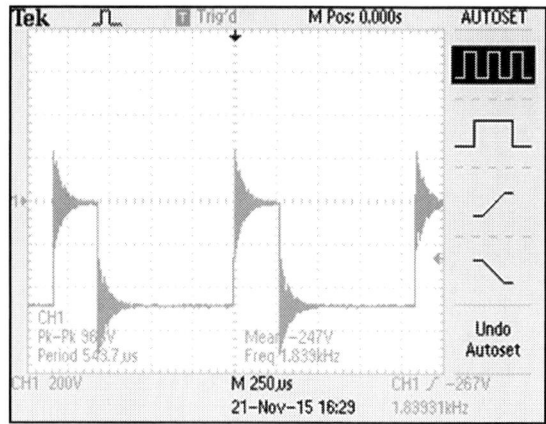

Fig 24: Boost voltage 9V to 24V using P&O Algorithm

Conclusion

Solar PV is a technology that offers a solution for a number of problems associated with fossil fuel. It is clean and continuously imports the energy from the sources. Round the globe India placed in the top which has highest solar irradiances. PV cell covert solar into electricity with a specified irradiance but due to partial shading irradiances goes lower and output power of the PV panel decreases. To get the better output Maximum Power Point (MPP) is track by designing different algorithm and we have designed controller based hardware using P&O algorithm of low cost and reduce complexity due to the automatic code generated in the MATLAB. Simulation and hardware result are successfully obtained.

Future Work

Designed low cost solar PV system can be installed in the small town, residential buildings etc.

The designed model can be used or implemented as a-

➢ Standalone unit.
➢ Design for the charge controller.
➢ Combination of the series and parallel connection can be explored.
➢ Cope up with the higher non linearity fuzzy algorithm to obtain MPP.
➢ For the larger PV system they can be implemented with the multi level inverter.
➢ Design of buck/boost algorithm can also be done to make the system reliable

References

[1] P.Kumar, D.K Palwalia "Decentralized Autonomous Hybrid Renewable Power Generation" "Hindawi Publishing Corporation" Volume 2015, Article ID 856075.
[2] R. Garg, A. Singh, S. Gupta "PV Models and Dynamic simulation of MPPT in MATLAB", IEEE Trans, 2014 pp 6-12.
[3] J.M. Blames, J.A. Carrasco, E. Avila "Maximum Power Point Estimator for Photovoltaic solar array" IEEE Mediterranean MELECON-2006 pp 889-892.
[4] J. Ahmad, "A Fractional open circuit voltage MPPT for PV Array" IEEE ,2nd ICSTE, 2010, Vol-1 pp 247-250.
[5] U. Kamran, S. Sirta, S. Yovsawat "A low cost Micro Controller based Maximum Power Point Tracking System with Multiple String connection for PV stand alone Application" IEEE 10th international conference, PEDS-13,pp 1343-1348.
[6] A. H. Niasar, Z. Zare, F.R. Far "A low cost P&O based MPPT ,combined with two degree sun tracking" IEEE, PEDSTC-2015,pp 119-124.
[7] H. A. Sher, AF Mautaza, A Noman "A new sensorless hybrid MPPT algorithm based on fractional short circuit current measurement and P&O MPPT" IEEE Transaction ,Vol 6,Issue 4 ,pp 1426-1434.
[8] B. Kuczeski, R. Philip, W. C Mensser "A platform for building PIC application for control and instrument" IEEE American conference Vol 7 pp 5162-5168.
[9] K. S. Phani Kiranmai, M. Veerachy "A single stage power conversion system for the PV MPPT application" IEEE international conference pp 2125-2130.
[10] G. Petrone, G. Spagavolo, M. Vitelli "An anolog techniques MPPT PV Application" IEEE Transaction Vol 59, Issue 12 PP 4713-4722.
[11] Kai Chen, Shulin Tain, Yahua Cheng, Libing Bai "An improved MPPT Controller for PV system under partial shading condition" IEEE Transaction Vol 5, Issue 3, PP 978-985.

Design and Implementation of Cost Effective Controller for Solar PV Application

[12] S. Bokswan, T. Benjanarasth "Code generation of fractional filter for DsPIC Micro controller" IEEE Conference, IENCON-2011 PP-1275-1279.

[13] F. Palmino, J. O Pinto, L. H Pereria "cost effective photovoltaic water pumping system for remote regions communities" IEEE conferences, ECCE-2014.PP-3426-3433.

[14] S. Ozturk, I.Cardirci "DSPIC Microcontroller based implementation of a fly back PV micro inverter using direct digital synthesis" IEEE conference, ECCE-2013, PP 3426-3433.

[15] A. Alnabusi, R. Dhauadi " Efficiency optimization of a DSP based standalone PV system using fuzzy logic an dual MPPT control" IEEE Transaction Vol 8, Issue 3, PP 573-584.

[16] laird, L. VDD "High step up dc dc topology and MPPT algorithm for use with a thermoelectric generator". IEEE Transaction PP 3147-3157.

[17] S.S. Valnjkar, S. D. Joshi, N. R. Kukarni "Implementation of MPPT charge controller for renewable energy" IEEE conference ICACCCT-2014, PP-255-259.

[18] Y. P. Siwakoti, B. B. Chhetri, D. Bista "Microcontroller based intelligent DC/DC converter to track maximum power point for solar photovoltaic module, IEEE conference CITRES 2010 PP 94-101.

[19] S. R. Osman, N. A. Rohin "Microcontroller based solar battery charging system with MPPT feature with low irradiance" IEEE conference CGAT 2013 PP 437-441.

[20] M. Killi, S. Samata "Modified perturb and observe MPPT algorithm for drift avoidance in photovoltaic system" IEEE Transactions Vol-62, Issue 9, PP 5549-5559.

[21] D. Sera, L. Mathe, T. Kerkes " on the P&O and INC MPPT methods for PV system" IEEE Journal Vol 3 Issue 3,PP 1070-1078.

[22] M. R. Islam, Guo Youguaugo, "Simulation of PV array characteristics and fabrication of Micro controller based MPPT" IEEE conference ICECE 2010, PP 155-158.

[23] M.S. Palli, F. M. Munshi, S. L. Medegar "Speed control of BLDC Motor with four switch three phase inverter using digital signal controller" IEEE conference, ICPACE-2015,PP-371-376.

[24] S. Z. Munji, R.A. Rahim, MHF Rahiman "Two controller interaction using C"IEEE conference 2010 PP 290-292.

[25] S. K. Kolimalla, M. K Mishra, "Variable perturbation size adaptive P&O ,MPPY Algorithm for sudden changes in irrandiance" IEEE transaction Vol 5 Issue 6 PP 718-728.

[26] H. Pate "MATLAB-Based Modeling to Study the Effects of Partial Shading on PV Array Characteristics" IEEE transaction Vol 23 Issue 1,2008 PP 302-310.

[27] T. Esram, P. L. Chapman "Comparison of Photovoltaic Array Maximum Power Point Tracking Techniques" IEEE transaction Vol 22 Issue 2, 2007 PP 439-449.

[28] P. Sanjeevi Kumar, G. Grandi "A Simple MPPT Algorithm for Novel PV Power Generation System by High Output Voltage DC-DC Boost Converter" IEEE Conference ISTE 2015 PP 214-220.

[29] Adithya Vangari, Divyanagalakshmi Haribabu "Modeling and Control of DC/DC Boost Converter using K-Factor Control for MPPT of Solar PV System" IEEE Conference ICEEE 2015 PP 1-6.

[30] Sarthak Jain, Anant Vaibhav "Comparative Analysis of MPPT Techniques for PV in Domestic Applications" IEEE Conference PIICON 2014, PP 1-6.

[31] Asim Datta, Dipankar Mukherjee, Sanjoy Debbarma "A DsPIC based Efficient Single-Stage Grid-Connected Photovoltaic System" IEEE conference TENCON 2014 PP 1-4.

[32] S. S. Valunjkar, S. D. Joshi "Implementation of Maximum Power Point Tracking Charge Controller for Renewable Energy" IEEE Conference ICACCCT 2014, PP 255-259.

[33] A. Nadjah, B. Kadri, "New Design of Dual Axis Sun Tracker with DSPIC Microcontroller" IEEE Conference PEMC-2014 PP 1030-1034.

[34] Mohammad B. Shadmand, Mostafa Mosa "An Improved MPPT Technique for High Gain DCDC Converter using Model Predictive Control for Photovoltaic Applications" IEEE Conference APEC 2014 PP 2993-2999.

[35] Ahmed Abdalrahman, Abdalhalim Zekry "Control of The Grid-connected inverter using DsPIC Microcontroller" IEEE conference JEC-ECC-2013, PP 159-164.

[36] Abdullah M. Noman, Khaled E. Addoweesh "Simulation and DSPACE Hardware Implementation of the MPPT Techniques Using Buck Boost Converter" IEEE conference CCECE 2014 PP 1-8.

[37] M. Elshaer, A. Mohamed "Smart Optimal Control of DC-DC Boost Converter in PV Systems" IEEE conference (T&DNA) -2015, PP 403-410.

[38] Lajos Török, Stig Munk-Nielsen "Simple Digital Country of a Two-Stage PFC Converter Using DSPIC30F Microprocessor" IEEE conference PEMD-2010 PP 1-4.

[39] Khanchai Tunlasakun, Krissanapong Kirtikara, Sirichai Thepa "Comparison Between FPGA-Based and DsPIC-Based Islanding Detection for Grid Connected Inverter in a Photovoltaic System" IEEE conferenceICCAS-2007 PP 416-419.

[40] T.-F. Wu, S.-A. Wang, C.-L. Kuo "Design and Implementation of a Push-Pull Phase-Shifted Bi-directional Inverter with a DsPIC Controller" IEEE conference PEDS-2009 PP 728-733.

[41] Nicola Femia, Giovanni Petrone "Optimization of Perturb and Observe Maximum Power Point Tracking Method" IEEE Trans. Vol 20 Issue 4 PP 963-973.

[42] Panom Petchjatupom, Wannaya Ngamkham "A Solar-powered Battery Charger with Neural. Network Maximum Power Point Tracking Implemented on a Low-Cost PlC-microcontroller" IEEE conference PEDS-2015 PP 507-510.

[43] K. S. Phani Kiranimai, M. Veerachary "A Single-Stage Power Conversion System for the PV MPPT Application" IEEE conference ICIT-2006 PP 2125-2130.

Code Generated For P&O Algorithm

```c
2    * PandO.c
3    *
4    * Code generation for model "PandO".
5    *
6    * Model version        : 1.4
7    * Simulink Coder version: 8.2 (R2012a) 29-Dec-2011
8    * C source code generated on: Mon Nov 30 13:30:34 2015
9    *
10   * Target selection: dspic.tlc
11   * embedded hardware selection: 16-bit Generic
12   * Code generation objectives: Unspecified
13   * Validation result: Not run
14   */
15   #include "PandO.h"
16   #include "PandO_private.h"
17
18   /* Real-time model */
19   RT_MODEL_PandO PandO_M_;
20   RT_MODEL_PandO *const PandO_M = &PandO_M_;
21   static void rate_monotonic_scheduler(void);
22
23   /*
24   * Set which subrates need to run this base step (base rate
always runs).
25   * this function must be called prior to calling the model
step function
26   * in order to "remember" which rates need to run this
base step. The
27   * buffering of events allows for overlapping preemption.
28   */
29   void PandO_SetEventsForThisBaseStep (boolean_T
*eventFlags)
30   {
31   /* Task runs when its counter is zero, computed via
rtmStepTask macro */
32   eventFlags[1] = ((boolean_T)rtmStepTask(PandO_M,
1));
33   }
34
35   /* rate_monotonic_scheduler */
36   static void rate_monotonic_scheduler (void)
37   {
38   /* Compute which subrates run during the next base time
step. Subrates
39   * are an integer multiple of the base rate counter.
Therefore, the subtask
40   * counter is reset when it reaches its limit (zero means
run).
41   */
42   (PandO_M->Timing.TaskCounters.TID [1]) ++;
43   if ((PandO_M->Timing.TaskCounters.TID [1]) > 1) {/*
Sample time: [0.002s, 0.0s] */
44   PandO_M->Timing.TaskCounters.TID [1] = 0;
45   }
46   }
47
48   /* Model output function for TID0 */
49   void PandO_output0(void)        /* Sample time:
[0.001s, 0.0s] */
50   {
51   {                              /* Sample time: [0.001s, 0.0s] */
52   rate_monotonic_scheduler();
53   }
54   }
55
56   /* Model update function for TID0 */
57   void PandO_update0(void)        /* Sample time:
[0.001s, 0.0s] */
58   {
59   /* (no update code required) */
60   }
61
62   /* Model output function for TID1 */
63   void PandO_output1(void)        /* Sample time:
[0.002s, 0.0s] */
64   {
65   /* (no output code required) */
66   }
67
68   /* Model update function for TID1 */
69   void PandO_update1 (void)        /* Sample time:
[0.002s, 0.0s] */
70   {
71   /* (no update code required) */
72   }
73
74   void Pando output (int_T tid)
75   {
76   switch (tid) {
77   case 0:
78   PandO_output0 ();
79   break;
80
81   case 1:
82   PandO_output1();
83   break;
84
85   default:
86   break;
87   }
88   }
89
90   void Pando update (int_T tid)
91   {
92   switch (tid) {
93   case 0:
94   PandO_update0();
95   break;
96
97   case 1 :  98   PandO_update1();
99   break;
100
101   default:
102   break;
103   }
104   }
105
106   /* Model initialize function */
107  voids Pando initialize (void)
108   {
109   /* Registration code */
110
111   /* initialize real-time model */
112   (void) memset((void *)PandO_M, 0,
```

```
113         sizeof(RT_MODEL_PandO));
114
115   /* S-Function "dsPIC_MASTER" initialization Block:
<S1>/Master */
116   /* Solver mode : Multitasking */
117   /* CONFIG TIMER 1 for scheduling steps        */
118   ConfigIntTimer1 (T1_INT_PRIOR_1 & T1_INT_ON);
119   T1CON = 0x8000;              /* T1_PS_1_1     */
120   PR1 = 4999;
121
122   /* Configuration TRIS */
123   /* Configuration ADCHS */
124   ADPCFG = 0U;
125 }
126
```

A New Approach of Offline Parameters Estimation for Vector Controlled Induction Motor Drive

Krishna R More*, P N Kapil** and Hormaz Amrolia***
* Department of Electrical Engineering, Institute of Technology, Nirma University, Ahmedabad, India
14meep17@nirmauni.ac.in
** Department of Electrical Engineering, Institute of Technology, Nirma University, Ahmedabad, India
pnkapil@nirmauni.ac.in
***Department of Electrical Engineering, Institute of Technology, Nirma University, Ahmedabad, India
hormaz.amrolia@nirmauni.ac.in

Abstract: For efficient performance of any vector controlled drive, motor parameters are to be estimated with utmost accuracy. Offline and online parameter estimation techniques are generally used for estimating parameters. In this paper, the offline technique is adopted due to its advantage over latter technique. Under offline parameter estimation, the new strategy is simulated for estimating the machine parameters. Inverter test and single phase test are used for estimating parameters. Only three switches of the inverter are used for estimating parameters in inverter test and four switches in single phase test. These tests are easy to implement with no extra hardware, no mechanical job is required and equations are easy to understand and implement. These tests are simulated in MATLAB Simulink and parameters obtained are compared with actual parameters of machine model available in MATLAB Simulink.

Keywords: Offline parameter estimation, inverter test, single phase test, dc resistance test, open circuit test, locked rotor test.

Introduction

Induction motors are widely used the electric machines in high-performance drive applications. Vector control of induction motor requires a precise knowledge of the motor parameters for its efficient operation. Numerous methods for estimating induction machine parameters such as online and offline parameter estimation have been developed exclusively for application in high-performance drives. Any vector controlled induction motor drive is inverter fed, numerous tests based on an inverter supply have been developed in recent past for determination of the required parameter values. Such methods are called offline parameter identification methods. Numerous possibilities exist nowadays to update the parameter values during the drive operation, termed as online parameter estimation methods [1], [2], [3]. Offline parameter estimation technique is adopted, as parameters are estimated at the initial stage itself with the inverter and then the estimated parameters are used in the system. There is no need to sense the signals continuously in offline parameter estimation, in order to estimate parameters as it is mandatory in online parameter estimation. No standard tests like an open circuit, locked rotor test are required where the mechanical task is required to estimate the parameters. Moreover, for estimating parameters offline, no PI controller is required. So, hardware requirement reduces. The only inverter, sensors, and controller are required in this technique. Offline parameter estimation technique uses inverter test and single phase test for estimating machine parameters. Inverter test for R_s, L_s and single phase test for L_{ls}, L_m and R_r.

Parameter Estimation Tests

An equivalent circuit of induction motor valid for steady state is shown in Fig. 1.The circuit is composed of two resistors, the stator resistance (R_s) and the rotor resistance (R_r), and three ideal inductors, the stator leakage inductance (L_{ls}), the rotor leakage inductance (L_{lr}) and the magnetizing inductance (L_m). The stator voltage is represented by V_s and s is the slip frequency normalized with respect to the stator frequency. The purpose of this work is to measure the values of these five ideal components.

Traditional Parameter Estimation Tests

DC Resistance Measurement

DC Measurement test is used to determine the stator resistance (R_s). This test consists of applying a DC voltage to stator winding. The DC voltage (V_{dc}) and current (I_{dc}) are measured and R_s value is computed by dividing both measurements. Basically, a dc voltage is applied to the stator windings of an induction motor. Because the current is dc, there is no induced

A New Approach of Offline Parameters Estimation for Vector Controlled Induction Motor Drive

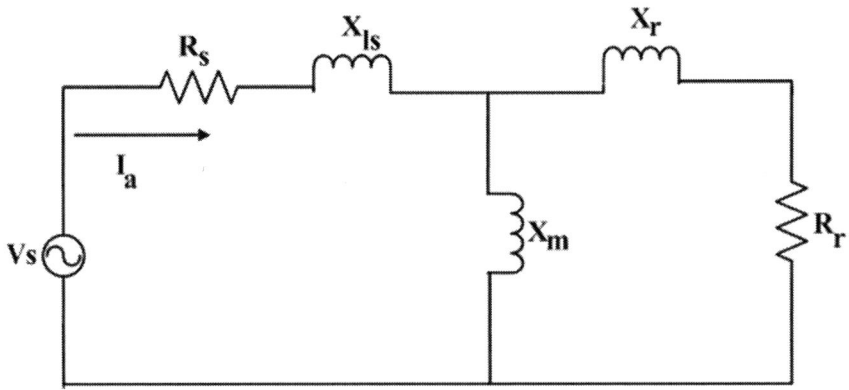

Figure 1: Steady State Induction Motor per phase equivalent circuit

voltage in the rotor circuit and no resulting rotor current. Also, the reactance of the motor is zero at direct current. Therefore, the only quantity limiting current now in the motor is the stator resistance, and that resistance can be determined. The basic circuit for the dc test is shown in Fig. 2 where, dc power supply connected to two of the three terminals of a Y-connected induction motor. [6]

$$R_s = \frac{V_{dc}}{2I_{dc}} \qquad (1)$$

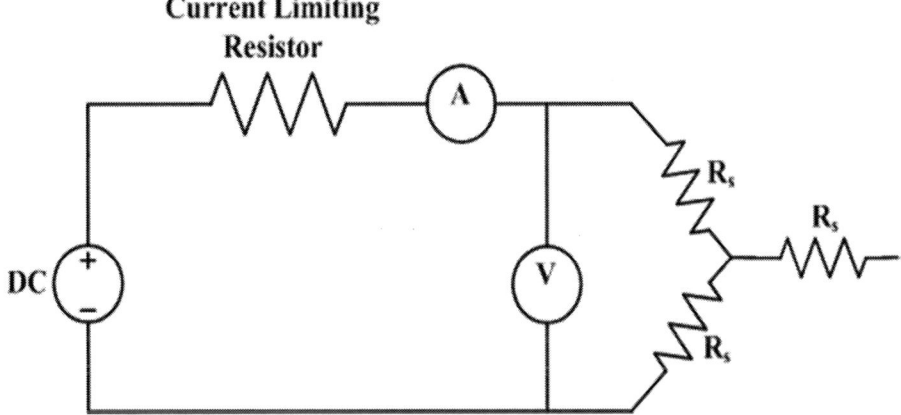

Figure 2: DC Resistance Test

No Load Test
Input power and motor current are measured at different voltage levels. Motor current and input power at rated voltage is called as no load current (I_{nl}) and no load power (P_{nl}) respectively. Motor no-load resistance, no load impedance, and no load reactance can be calculated. No load test is performed to calculate no load (stator) reactance (X_{nl}) or stator inductance (L_s). The motor is allowed to run at rated voltage and allowed to reach rated speed on no-load. The equation to measure (X_{nl}) is given below (equ.2). The equivalent circuit of induction motor under a no-load condition is shown below in Fig. 3. The simulation arrangement for open circuit test is also shown below in Fig. 4 [5].

$$X_{nl} = \frac{Q}{I^2} \qquad (2)$$

$$Q = \sqrt{(V_a I_a)^2 + (P_a)^2} \qquad (3)$$

where, Q = Reactive Power, V_a = Phase Voltage, I_a = Phase Current, P_a = Active power per phase, X_{nl} = No-Load Reactance

Seventh International Conference on Advances in Power Electronics and Instrumentation Engineering – PEIE 2016

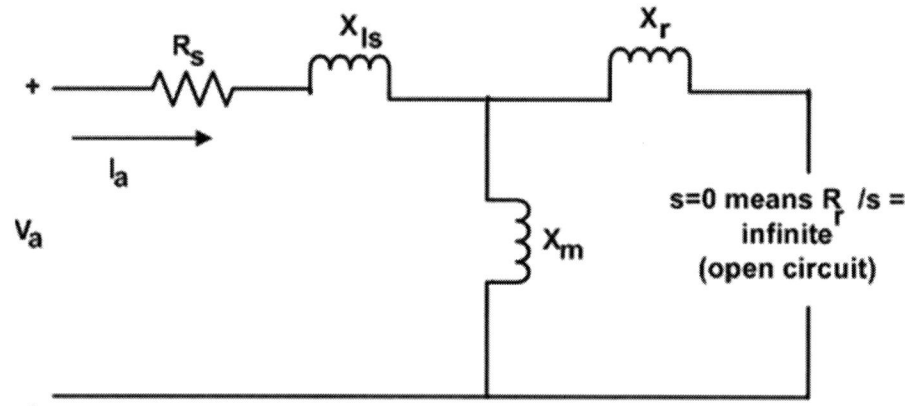

Figure 3: Three phase equivalent circuit of induction motor under no load condition

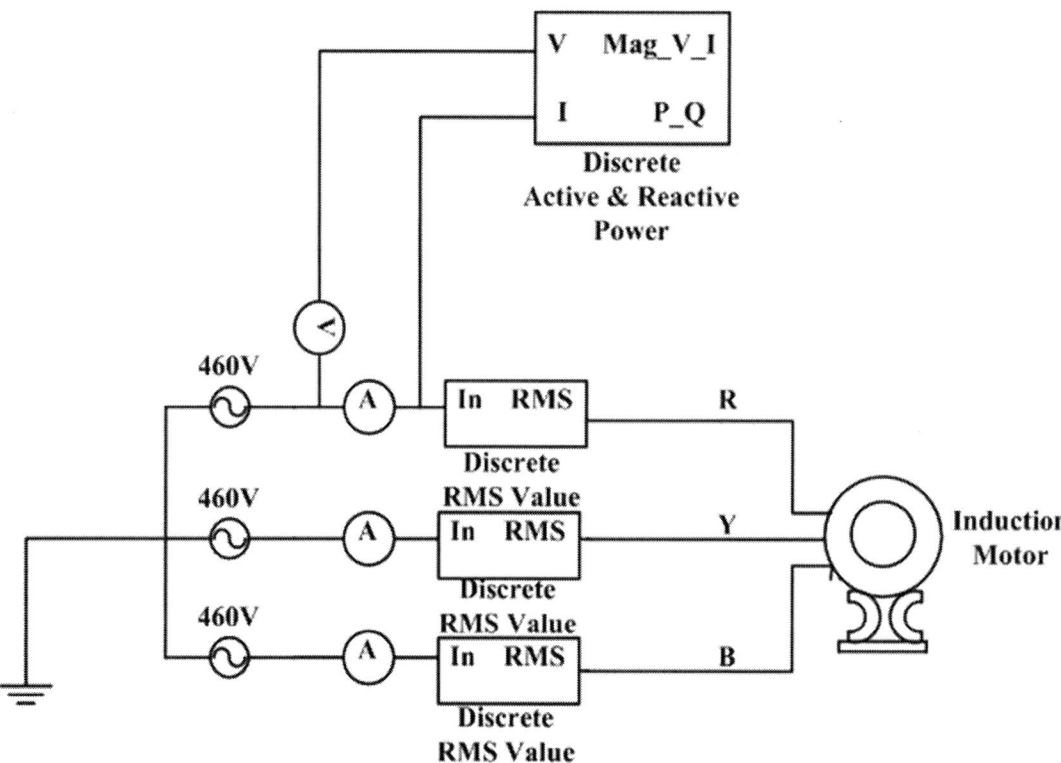

Figure 4: Simulation arrangement for induction motor under no load condition

Locked Rotor Test

Motor phase to phase voltage and input power at rated current are called as locked rotor voltage (V_{lr}) and locked rotor power (P_{lr}). Motor locked rotor resistance (R_r), locked rotor impedance (Z_r) and locked rotor reactance (X_r) can be calculated. Fig.5 shows the behavior of an induction motor under locked rotor test. Locked rotor test is performed to obtain stator leakage reactance (X_{ls}) (equ.4) basically stator leakage inductance (L_{ls}) is estimated (equ.5), magnetizing reactance (X_m) i.e. magnetizing inductance (L_m) can be estimated from (equ.6) and (equ.7) respectively. Rotor resistance (R_r) is estimated by using X_{ls}, X_m and R_2 (equ.11). In this test, the motor is allowed to run at a voltage less than the rated voltage till the rated current flows through the windings. Another way to estimate Rr is by using power factor (equ.12) and (equ.13). The simulation arrangement for locked rotor test is also shown below Fig.6 [5].

20

A New Approach of Offline Parameters Estimation for Vector Controlled Induction Motor Drive

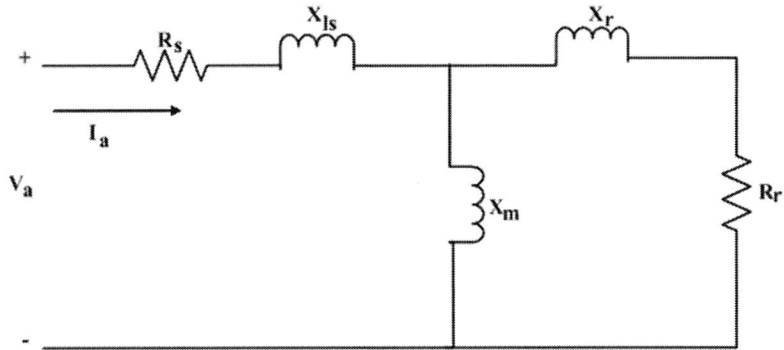

Figure 5: Simplified equivalent circuit of three phase induction motor under locked rotor test

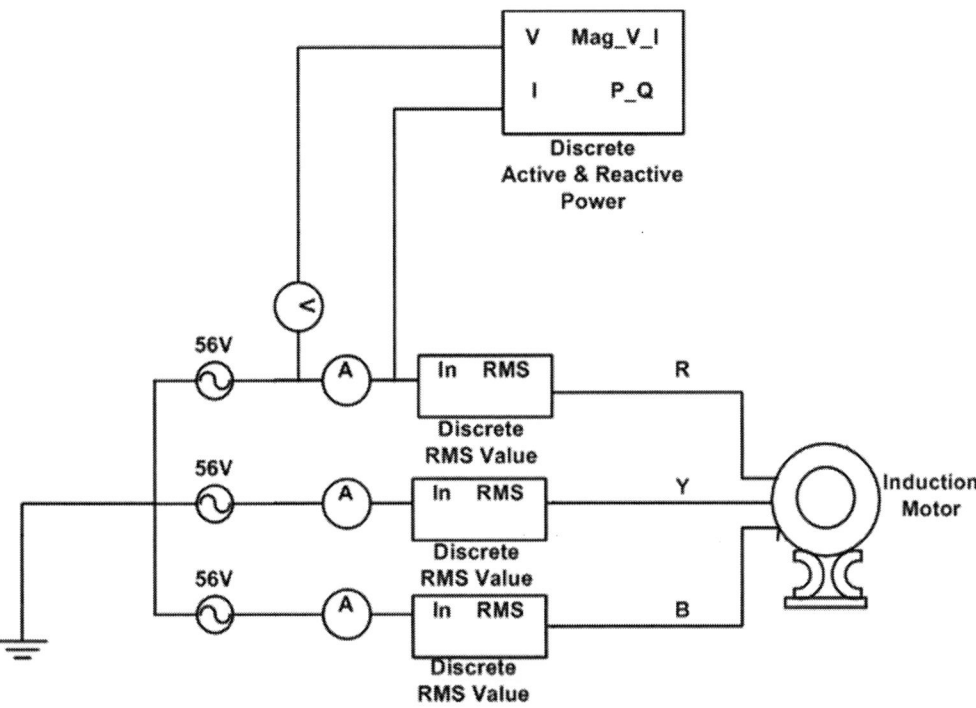

Figure 6: Simulation arrangement for induction motor under locked rotor condition

The equations for estimating various parameters are mentioned below:-

$$X_{ls} = 0.3 * \frac{Q}{I^2} \qquad (4)$$

where, Q is Reactive Power
Here, 0.3 factor to be multiplied for the class C motor [4].

$$L_{ls} = L_{lr} = \frac{X_{ls}}{2\pi f} = \frac{X_{lr}}{2\pi f} \qquad (5)$$

$$X_m = X_{nl} - X_{ls} \qquad (6)$$

$$L_m = \frac{X_m}{2\pi f} \qquad (7)$$

$$L_s = L_{ls} + L_m \qquad (8)$$

Seventh International Conference on Advances in Power Electronics and Instrumentation Engineering – PEIE 2016

$$R = \frac{P}{I^2} \tag{9}$$

where, P is Active Power

$$R_2 = R - R_s \tag{10}$$

$$R_r = R_2 * \left(\frac{X_{ls} + X_m}{X_m}\right)^2 \tag{11}$$

$$\cos\phi_s = \frac{P}{\sqrt{3} * V_a * I} \tag{12}$$

$$V_a = \left(I_{sc} \angle \phi_s\right) * \left(R_s + R_r + jX_{ls}\right) \tag{13}$$

where, I_{sc} is short circuit current and ϕ_s is lagging power factor angle.

The calculation results are compared and percentage error between the actual parameter of machine model available in MATLAB Simulink and calculated parameter through above equations is also calculated are shown in tables below.

Table 1. For 5 HP Machine Model available in MATLAB

Parameter	Calculated Value	Actual Value	Percentage Error (%)
R_s	0.923 Ω	1.115 Ω	17.21
L_s	0.20895 H	0.209674 H	0.345
L_{ls}	0.00355 H	0.005974 H	40.57
L_m	0.2054 H	0.2037 H	0.834
R_r	1.0689 Ω	1.083 Ω	1.3

Table 2. For 10 HP Machine Model available in MATLAB

Parameter	Calculated Value	Actual Value	Percentage Error (%)
R_s	0.5851 Ω	0.6837 Ω	14.42
L_s	0.15226 H	0.152752 H	0.3
L_{ls}	0.00246 H	0.004152 H	40.75
L_m	0.1498 H	0.1486 H	0.8
R_r	0.4956 Ω	0.451 Ω	1.2

Table 3. For Default HP Machine Model available in MATLAB

Parameter	Calculated Value	Actual Value	Percentage Error (%)
R_s	0.02873 Ω	0.029 Ω	0.93
L_s	0.035175 H	0.035489 H	0.88
L_{ls}	0.000355 H	0.000599 H	40.73
L_m	0.03482 H	0.03459 H	0.66
R_r	0.02052 Ω	0.22 Ω	6.81

Table 4. For 1 HP Machine Model available in MATLAB

Parameter	Calculated Value	Actual Value	Percentage Error (%)
R_s	9.589 Ω	12.625 Ω	24.05
L_s	0.803 H	0.8117 H	0.07
L_{ls}	0.0361 H	0.0617 H	41.49
L_m	0.7762 H	0.750 H	3.49
R_r	11.63 Ω	12.316 Ω	5.57

Proposed Parameter Estimation Methods

Buck Chopper

Fig.7 shows the circuit diagram of buck chopper. When the switch is ON energy is supplied to the load and is simultaneously stored in the inductor. During the ON period diode is in reverse bias condition. When the switch is OFF the energy stored in the inductor is fed to the load, which flows through the diode making it forward bias. Thus, the load is supplied with energy at every instant as long as input supplies the energy. Duty ratio decides the ON and OFF period of the switch.

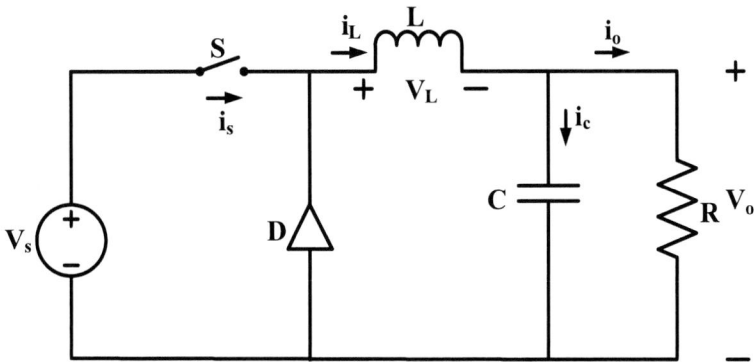

Figure 7: Buck Chopper

Here, in buck chopper 75% is set as duty ratio for 1 kHz frequency at 10 V. The value of L=10 mH and load is 10Ω. The voltage and current waveforms are shown in Fig.8 and Fig.9.

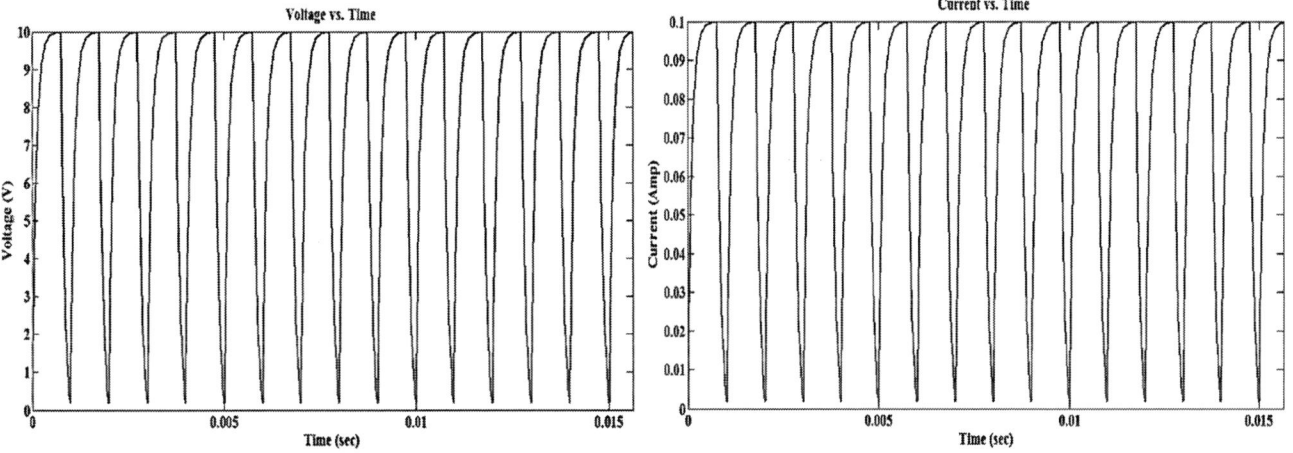

Figure 8: Voltage Waveform   Figure 9: Current Waveform

Inverter as a Buck Chopper

A new strategy is proposed for measuring R_s and L_s using an inverter, which is operated as a buck chopper. Fig.10 shows inverter operating as a buck chopper. As shown in Fig. 10, when switch S1 and S6 are ON, current flows through motor windings through phase R and phase Y and circuit gets complete as S6 is ON. When S1 is OFF energy stored in motor windings flows through phase Y and phase R as S6 and S4 are ON now as shown in Fig. 11. The duty ratio is 75% for S1 and it is 99.9% for S6. Only three switches S1, S4, S6 are being turned on to estimate these parameters. The R-phase and Y-phase are only in action i.e. the machine is getting supply from R and Y phase only. The value of L_s is calculated by observing the value of time at 63.8% of I_{max} (definition of the time constant). The magnified image of the current waveform is shown in the Fig.14, which shows the value of 63.8% of I_{max} is obtained at 0.1087 sec.

Seventh International Conference on Advances in Power Electronics and Instrumentation Engineering – PEIE 2016

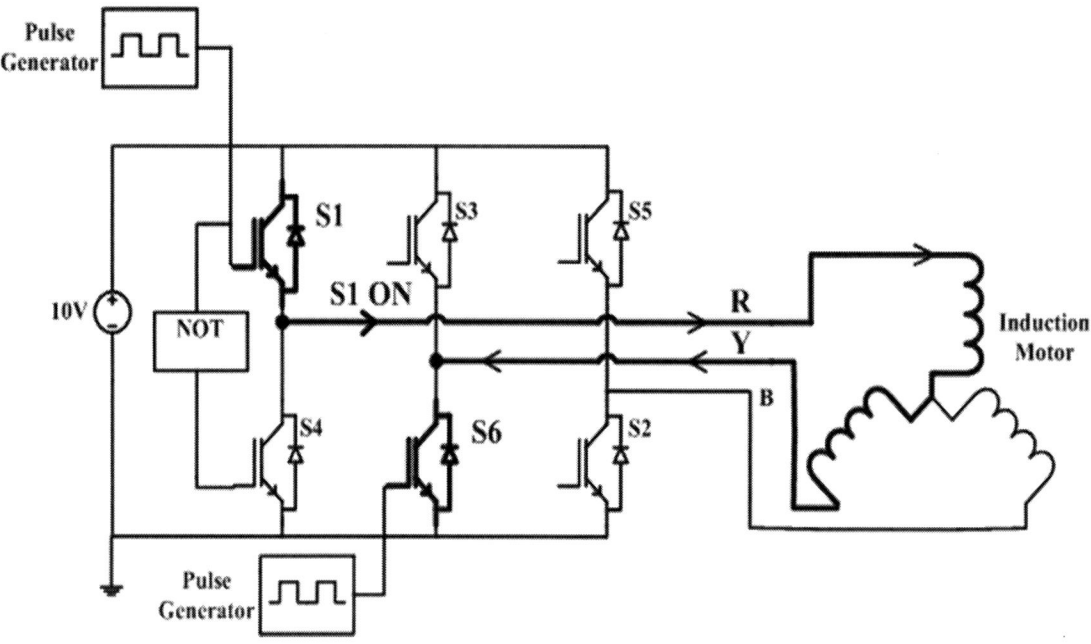

Figure 10: Inverter Circuit for estimating R_s & L_s - ON State

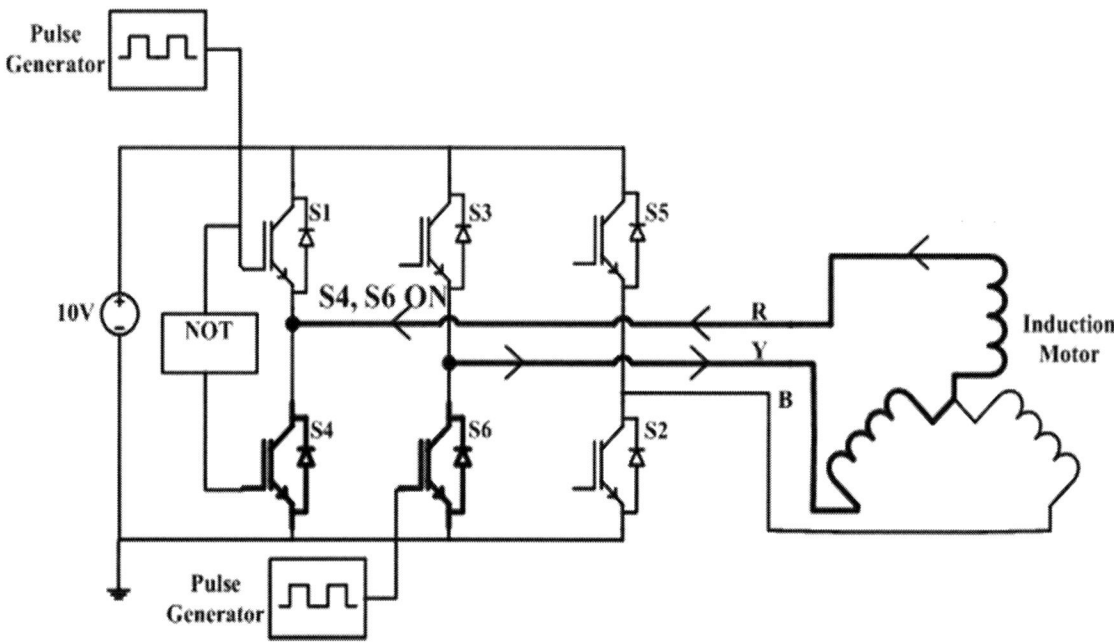

Figure 11: Inverter Circuit for estimating R_s & L_s - OFF State

The R_s is measured with the formula (equ.14) and the time constant is used to measure L_s (equ.15) with the help of current waveform.

$$R_s = \frac{DV_{dc}}{2I_{dc}}$$

(14)

where, D = Duty Ratio of the switch

$$\tau = \frac{L_s}{R_s}$$

(15)

where, τ = Time constant i.e. Time at 63.8% of Imax
The voltage and current waveform are shown in the figures below.

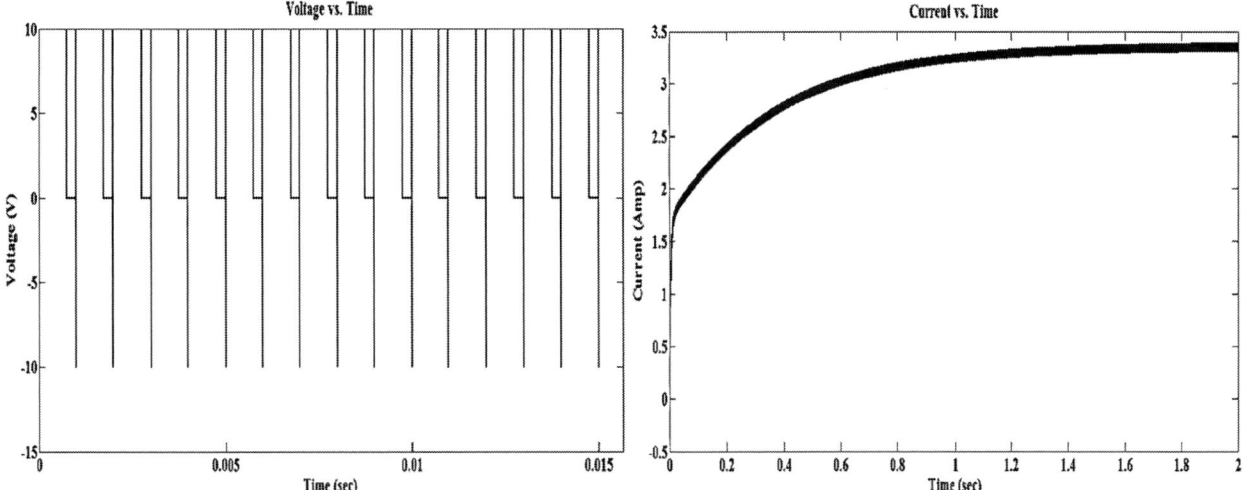

Figure 12: Voltage Waveform   Figure 13: Current Waveform

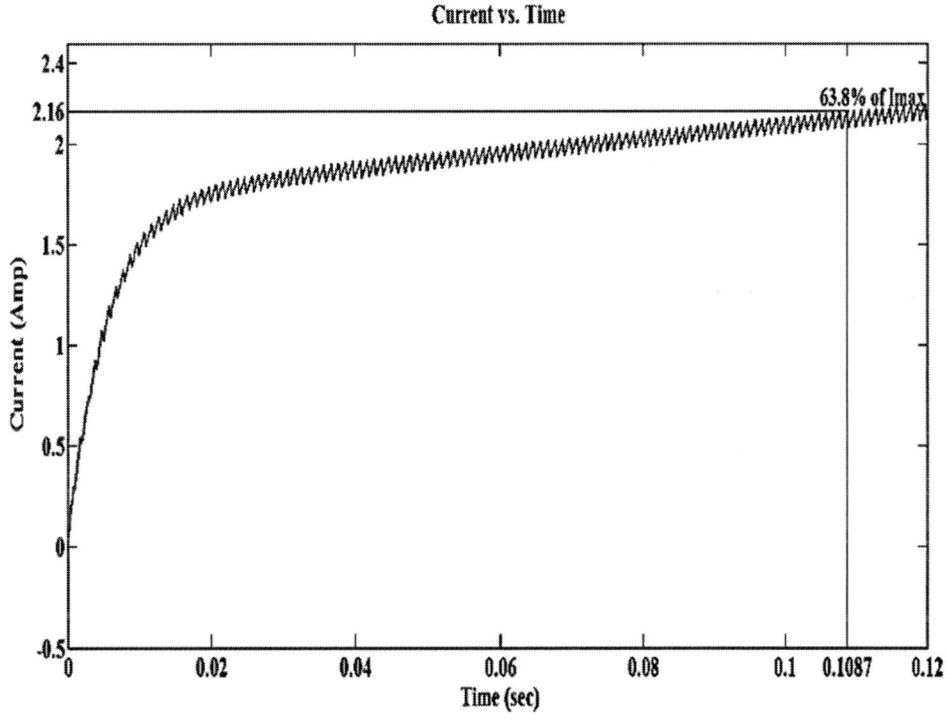

Figure 14: Magnified Current Waveform

Single Phase Test as Locked Rotor test
In single phase test motor is supplied with on only single phase by disconnecting one of the phases out of the three phases. Thus, the condition becomes like locked rotor test, where the rotor is blocked manually. So, equations used for calculating the parameter in locked rotor tests can be used to estimate parameters through single phase test. This test is used to estimate stator leakage inductance (L_{ls}), magnetizing inductance (L_m), rotor resistance (R_r). The circuit diagram for single phase test is shown in figures below. The current flows through S1, S6 for the positive half cycle Fig.15 and for negative half cycle it flows through S3, S4 Fig.16. The voltage and current waveform are shown in Fig.17 and Fig.18 respectively.

Seventh International Conference on Advances in Power Electronics and Instrumentation Engineering – PEIE 2016

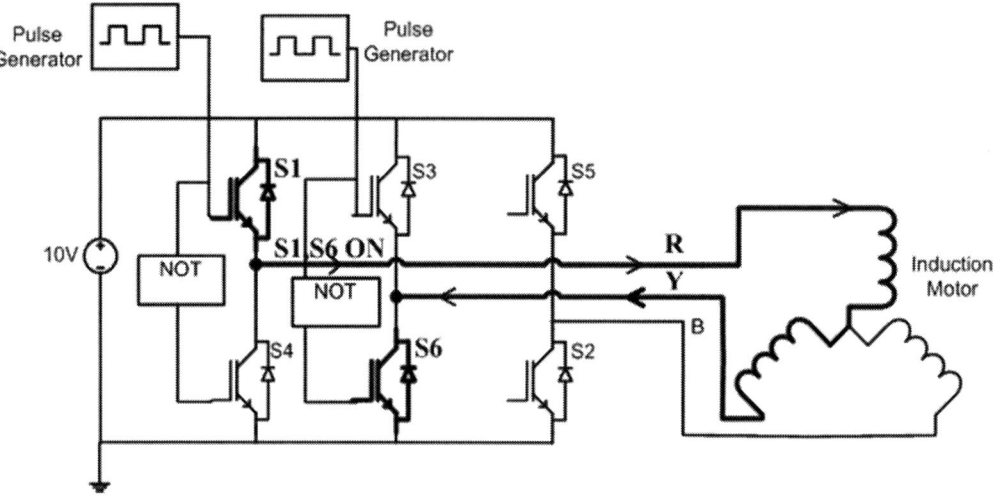

Figure 15: Inverter Circuit for Single Phase test - ON State

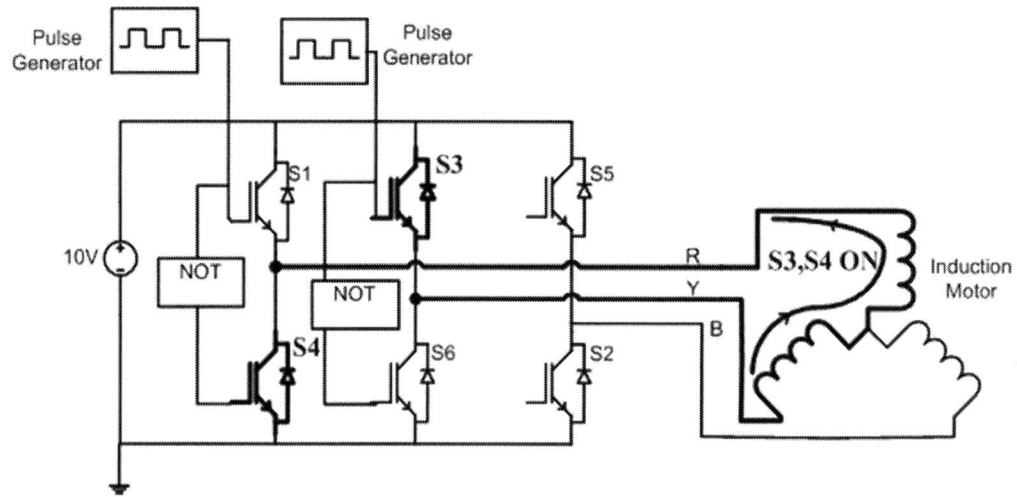

Figure 16: Inverter Circuit for Single Phase test - OFF state

Figure 17: Voltage Waveform

Figure 18: Current Waveform

26

A New Approach of Offline Parameters Estimation for Vector Controlled Induction Motor Drive

The calculation results are compared and percentage error between the actual parameter of machine model available in MATLAB and calculated parameter through equations is also calculated are shown in tables below.

Table 5. For 5 HP Machine Model available in MATLAB

Parameter	Calculated Value	Actual Value	Percentage Error (%)
R_s	1.105 Ω	1.115 Ω	0.9
L_s	0.12 H	0.209674 H	42.768
L_{ls}	0.0050 H	0.005974 H	16.308
L_m	0.115 H	0.2037 H	43.54
R_r	2.115 Ω	1.083 Ω	95.3

Table 6. For 10 HP Machine Model available in MATLAB

Parameter	Calculated Value	Actual Value	Percentage Error (%)
R_s	0.6787 Ω	0.6837 Ω	0.73
L_s	0.02898 H	0..152752 H	81.03
L_{ls}	0.00333 H	0.004152 H	19.80
L_m	0.02565 H	0.1486 H	82.74
R_r	1.35 Ω	0.451 Ω	199.33

Table 7. For Default HP Machine Model available in MATLAB

Parameter	Calculated Value	Actual Value	Percentage Error (%)
R_s	0.0299 Ω	0.029 Ω	3.10
L_s	0.0123 H	0.035489 H	65.34
L_{ls}	0.000457 H	0.000599 H	23.71
L_m	0.011843 H	0.03459 H	65.76
R_r	0.0684 Ω	0.022 Ω	210.91

Table 8. For 1 HP Machine Model available in MATLAB

Parameter	Calculated Value	Actual Value	Percentage Error (%)
R_s	12.5 Ω	12.625 Ω	0.1
L_s	0.46 H	0.8117 H	43.33
L_{ls}	0.0516 H	0.0617 H	16.37
L_m	0.4084 H	0.750 H	45.55
R_r	21.82 Ω	12.316 Ω	77.17

Observations
i. With the increase in power of motor the percentage error in R_s decreases while it increases for L_s for the inverter test.
ii. For a specific motor at a constant frequency but varying duty ratio (increasing) percentage error in R_s decreases and also for L_s it decreases.
iii. For a specific motor at constant duty ratio but increasing frequency, percentage error decreases in Rs and Ls.
iv. From calculations, it is found that the value of R_r is less than R_s.
v. From calculation, it is found that the percentage error in L_{ls} is more but the combined percentage error of L_{ls} and L_m i.e. L_s is less i.e. actual L_s equivalent to calculated L_s.
vi. It is found that percentage error of L_s calculated from inverter test is more than that found from open circuit test.
vii. It is found that percentage error of R_r calculated considering power factor is more compared to that when power is not considered.
viii. It is concluded that the percentage error between the actual parameters and calculated parameters is more in single phase test as compared to the lock rotor test.

Conclusion
In offline parameter estimation technique, inverter test, and single phasing test are simulated in MATLAB Simulink for estimating induction machine parameters. Parameters are also estimated with standard tests. The estimated parameter value is compared with an actual parameter value of machine model from MATLAB Simulink. The percentage error is calculated for each parameter. It is observed that R_s is estimated more accurately with inverter test in comparison with DC resistance test. Estimation of L_s is quite accurately done with inverter test. L_{ls}, L_m, R_r are quite accurately estimated from single phase test. When compared with parameter estimation technique used in [1], parameters are precisely estimated with inverter test and single phase test. Hence, with less hardware requirement and complexity, parameters are precisely estimated with inverter test and single phase test.

Seventh International Conference on Advances in Power Electronics and Instrumentation Engineering – PEIE 2016

References

[1] Jun Zheng, Yunkuan Wang, Xiaofei Qin and Xin Zhang, "An Offline Parameter Identification Method of Induction Motor", Proceedings of the 7th World Congress on Intelligent Control and Automation, Year: 2008, Pages 8898-8901.

[2] Hamid Toliyat, Emil Levi, Mona Rana "A Review of RFO Induction Motor Parameter Estimation Techniques ", Power Engineering Review, IEEE, Volume: 22, Year: 2002, Pages: 271-283.

[3] Hubert Schierling, "Fast And Reliable Commissioning of AC Variable Speed Drives By Self-Commissioning ", Industry Applications Society Annual Meeting, 1988, Conference Record of the 1988 IEEE, Volume: 1, Year: 1988 Pages: 489 - 492.

[4] K.K.Pandey, P. H. Zope "Estimating Parameters of a Three-Phase induction motor using Matlab/Simulink ", International Journal of Scientific and Engineering Research, Volume: 4, Year: 2013, Pages: 425 – 431.

[5] Saffet Ayasun, Chika O. Nwankpa "Induction Motor Tests Using MATLAB/Simulink and Their Integration into Undergraduate Electric Machinery Courses ", IEEE Transactions on Education Volume: 48, Year: 2005, Pages: 37 - 48.

[6] http://iitg.vlab.co.in

A Comparative Study of Switching Strategies for Single Phase Matrix Converter

Mohammadamin Yusufji Khatri * and Hormaz Amrolia **
*Department of Electrical Engineering, Institute of Technology, Nirma University, Ahmedabad, India
aminykhatri@hotmail.com
** Department of Electrical Engineering, Institute of Technology, Nirma University, Ahmedabad, India
hormaz.amrolia@nirmauni.ac.in

Abstract: AC-AC power conversion particularly for speed control of AC drives is done with single phase cyclo-converters. In this work the single-phase matrix converter (SPMC) topology is used as a cyclo-converter. IGBT are used as power switches. The sinusoidal pulse width modulation has been used for generation of pulses. This paper presents three different switching strategies for SPMC. A comparison is made between these three strategies based on parameters such as type of load to which the strategy is restricted, number of switches kept ON during single time interval, output voltage waveform of SPMC and its THD.

Keywords: Sinusoidal Pulse Width Modulation (SPWM), Insulated Gate Bipolar Transistor (IGBT), Single Phase Matrix Converter (SPMC), Pulse Width Modulation (PWM), Complementary of PWM (cPWM).

Introduction

AC-AC conversion consist of converting fixed ac voltage with fixed frequency to variable ac voltage with variable frequency, which can be done by two methods: 1) Indirect method and 2) direct method. Indirect method is the most common approach for the ac-ac power conversion which consist of a rectifier at supply side and inverter at the load side. Such an arrangement would require energy storage element like capacitor or an inductor in the intermediate dc link. These element make the converter bulky and useless in application requiring regenerative operation. These limitation can be overcome by the direct method for ac-ac conversion without any intermediate dc link [1]. Such an operation is performed by a cyclo-converter which converts the ac power at one frequency directly to another frequency. The most desirable features for any power frequency changer are: 1) Simple and compact power circuit 2) Generation of load voltage with arbitrary amplitude and frequency 3) Sinusoidal input and output currents 4) Operation with unity power factor for any load 5) Regeneration capability. These characteristics are not fulfilled by the conventional cyclo-converters, which leads to the use of matrix converter topology as it fulfills the ideal features. The matrix converter provides an "all silicon" solution for direct ac-ac conversion without any intermediate dc link thus eliminating the use of reactive energy storage elements. It consists of bidirectional switches which allows any output phase to be connected to any input phase. The topology was first proposed by Gyugyi. In recent years, matrix converter has gained a lot of attention for the traction application. The commutation between the switches in a matrix converter results in current spikes which are their major drawback [2]. Development of three phase matrix converter started with the work of Venturini and Alesina published in 1980 [4]. The SPMC was first developed by Zuckerberger [5]. Study of other SPMC topology had been carried out by Hossieni [6] and Abdollah Khoei [7] and Saiful [8]. Due to the absence of natural free-wheeling paths commutation issues need to be resolved in any PWM type of converter [9]. When inductive loads are used there may be switching spikes [10]. Switching arrangements for safe commutation is proposed by Zahiruddin. [11]. Amongst previous development of the cycloconverter includes; work on improvements of harmonic spectrum in the output voltage with new control strategies [12], new topology [13] and study of the cyclo-converter behaviour [14]. In this work, three different switching strategies are studied for obtaining AC-AC conversion from 50Hz to 150Hz. A computer simulation model on SPMC for cyclo-converter operation using MATLAB/Simulink (MLS) software package is developed. The simulation results for all three strategies have been portrayed and studied in this paper with all their advantages and limitation.

The single phase matrix converter as a cyclo-converter

The SPMC requires four bi-directional switches capable of blocking voltage and conducting current in both directions for its cyclo-converter operation. The basic circuit diagram is shown in Figure. 1. Currently due to the unavailability of any such discrete semiconductor device which could fulfil the needs, hence Common Emitter anti-parallel IGBT, diode pair is used as shown in Figure. 2. Diodes are used to provide reverse blocking capability to the switch module. IGBT's are used due to its

Seventh International Conference on Advances in Power Electronics and Instrumentation Engineering – PEIE 2016

high switching capabilities and high current carrying capacities for high power applications

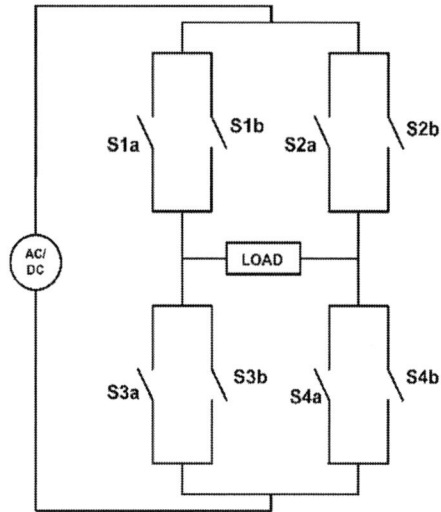

Figure 1: Basic circuit diagram of SPMC   Figure 2: Bidirectional switch

Switching strategies for SPMC as cyclo-converter

Strategy I

In this strategy only two switches are kept on at a time. One of the two switch is provided with a PWM pulse and other switch is modulated with a continuous pulse. The switching sequence is shown in Table 1.

- At any time 't' only two switches S1a and S4a will be kept ON and conduct the current flow during the positive cycle of input source (state 1). (Ref Figure 3)
- At any time 't' only two switches S1b and S4b will be kept ON and conduct the current flow during the negative cycle of input source (state 2). (Ref Figure 4)
- At any time 't' only two switches S2a and S3a will be kept ON and conduct the current flow during the positive cycle of input source (state 3). (Ref Figure 5)
- At any time 't' only two switches S2b and S3b will be kept ON and conduct the current flow during the negative cycle of input source (state 4). (Ref Figure 6)

The switching signals according to Table 1 is shown in Figure 7. Due to the turn off characteristics of IGBT the practical realization of the switching sequence in the SPMC is not instantaneous and simultaneous. Here the tailing off of the collector current will create a short circuit with the next switch turn ON especially when the inductive loads are used, resulting in switching spikes [16]. This strategy leads to two damaging effects which would lead to undue stress and destruction of switches: 1) current spikes are generated in the short circuit paths 2) voltage spikes will be induced due to change in current across inductance of load.

Table 1: Switching sequence for strategy I

Input frequency	Target output frequency	Time Interval	State	Switch modulated	PWM switch
50 Hz	150 Hz	1	1	S1a	S4a
		2	3	S2a	S3a
		3	1	S1a	S4a
		4	2	S1b	S4b
		5	4	S2b	S3b
		6	2	S1b	S4b

A Comparative Study of Switching Strategies for Single Phase Matrix Converter

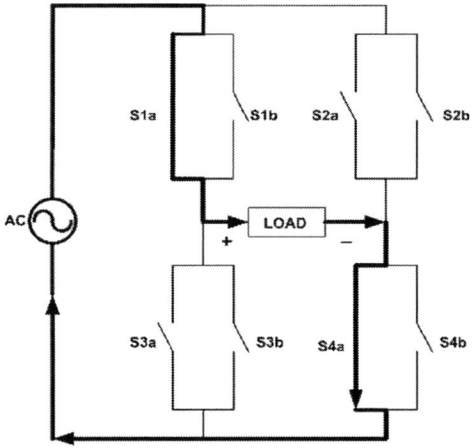

Figure 3: State 1 positive cycle

Figure 4: State 2 negative cycle

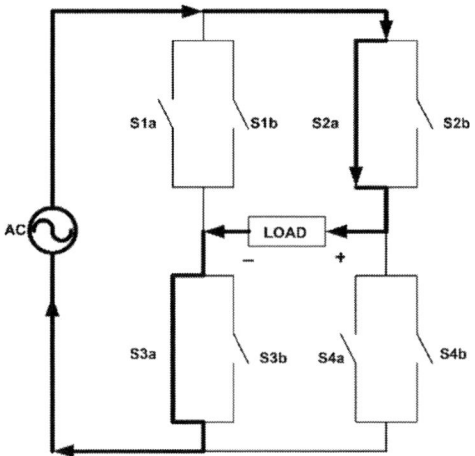

Figure 5: State 3 positive cycle

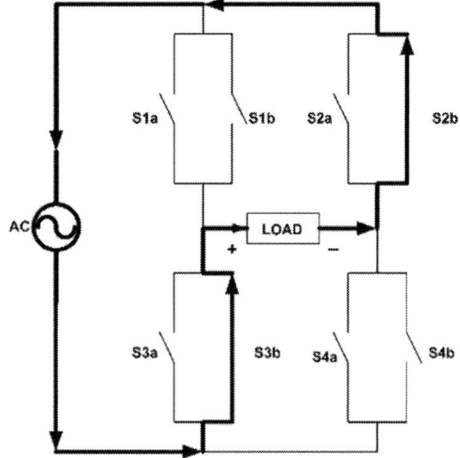

Figure 6: State 4 negative cycle

Both will subject the switches with undue stress leading to its destruction. Thus it cannot be used with RL load as it provides no path to discharge the energy across the inductor. Switching arrangements for safe-commutation as proposed by Zahirrudin can be used to eliminate the generation of switching spikes [17]. Refer Figure 8 to Figure 15 for the output voltage, current and their THD. Strategy II overcomes this limitation.

Strategy II
In this strategy three switches are used for a single interval out of which one is a PWM pulse and other two are modulated with a continuous pulse. The switching sequence are shown in the Table 2. Refer Figure 24 to Figure 27 respectively to understand the different operating states shown below.
- At any time 't', two switches S1a and S4a (PWM) will be kept ON and conduct the current flow during the positive cycle of input source, with S2b turn 'ON' for commutation purpose (state 1).
- At any time 't', two switches S4b and S1b (PWM) will be kept ON and conduct the current flow during the negative cycle of input source, with S3a turn 'ON' for commutation purpose (state 2)
- At any time 't', two switches S2a and S3a (PWM) will be kept ON and conduct the current flow during the positive cycle of input source, with S1b turn 'ON' for commutation purpose (state 3)
- At any time 't', two switches S3b and S2b (PWM) will be kept ON and conduct the current flow during the negative cycle of input source, with S4a turn 'ON' for commutation purpose (state 4)

This scheme can be used with R as well as RL load and is thus better than previous one. The THD can be further improved with proposed switching strategy. Figure 16 shows the switching signals according to Table 2. Refer Figure 17 – Figure 20 for the output voltage, current waveforms and their THD.

31

Seventh International Conference on Advances in Power Electronics and Instrumentation Engineering – PEIE 2016

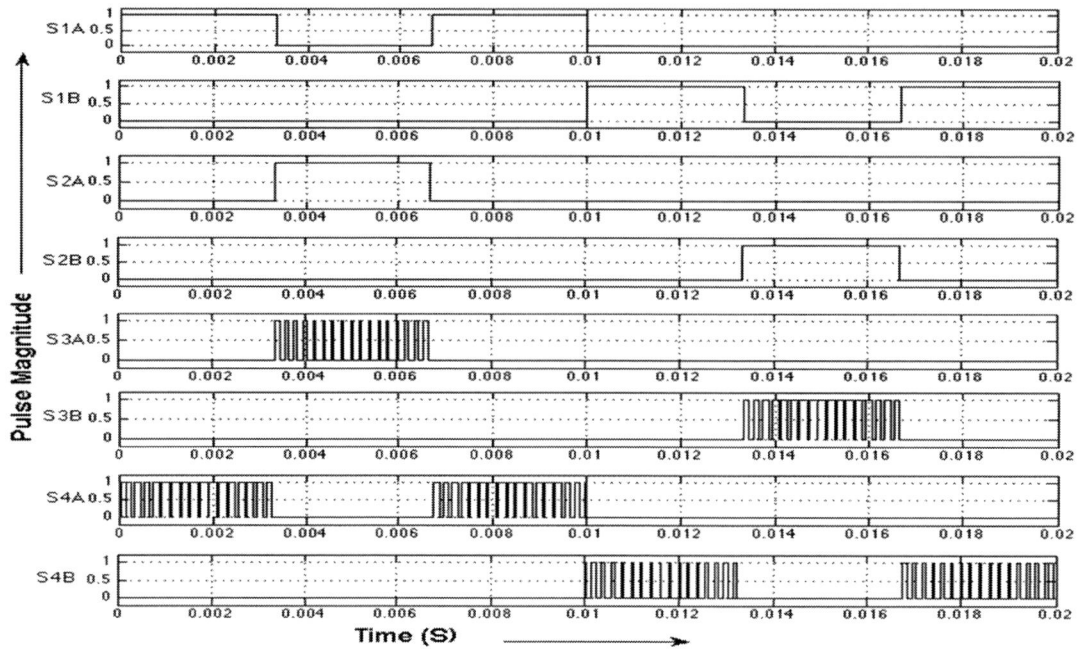

Figure 7: Switching signals for strategy I

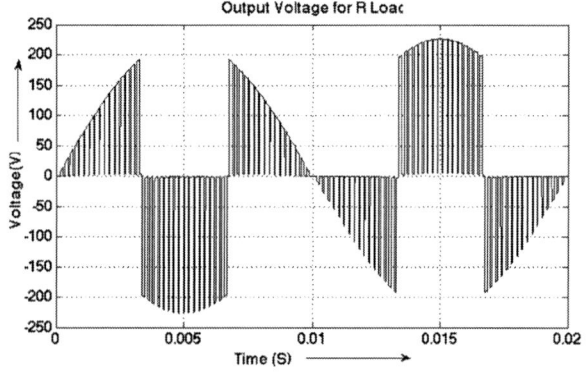

Figure 8: Output voltage for strategy I (R load)

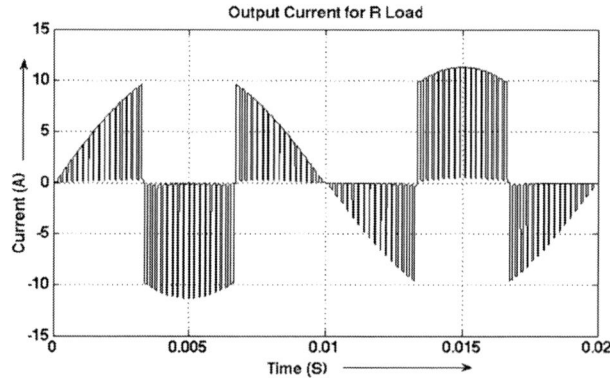

Figure 9: Output current for strategy I (R load)

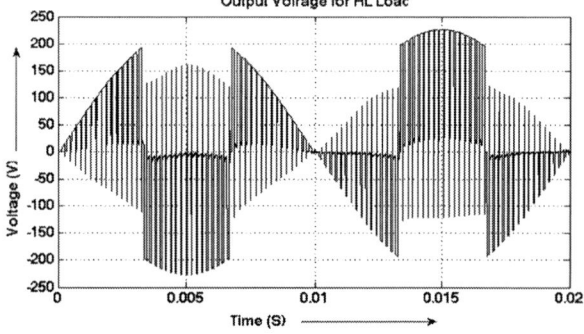

Figure 10: Output voltage for strategy I (RL load)

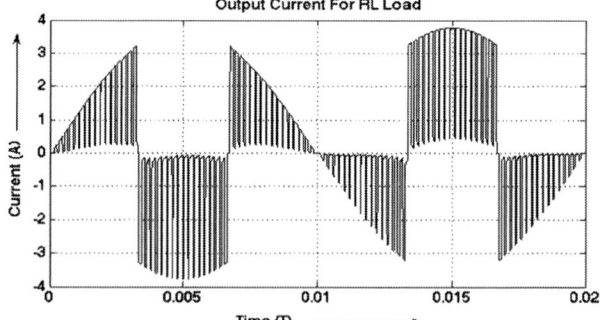

Figure 11: Output current for strategy I (RL load)

A Comparative Study of Switching Strategies for Single Phase Matrix Converter

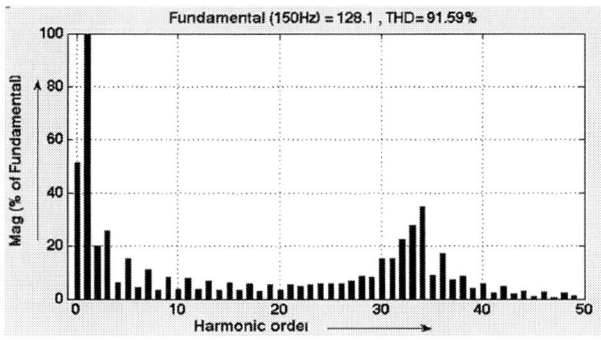

Figure 12: FFT analysis for output current with R load

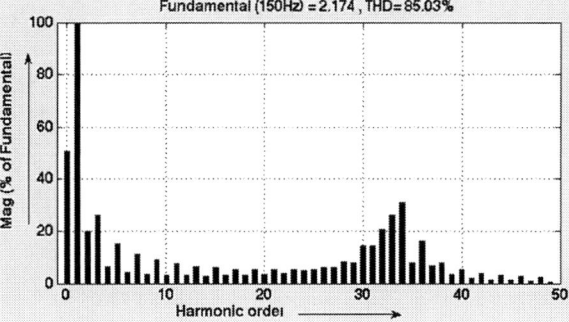

Figure 13: FFT analysis for output current with RL load

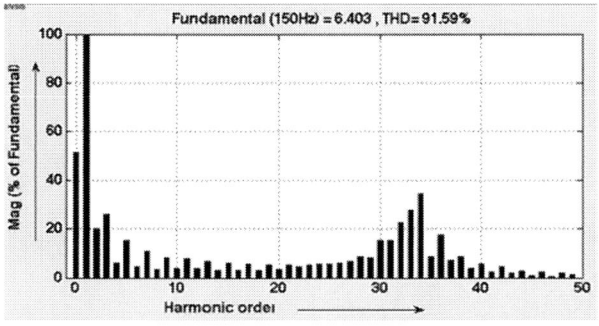

Figure 14 FFT analysis for output voltage with R load

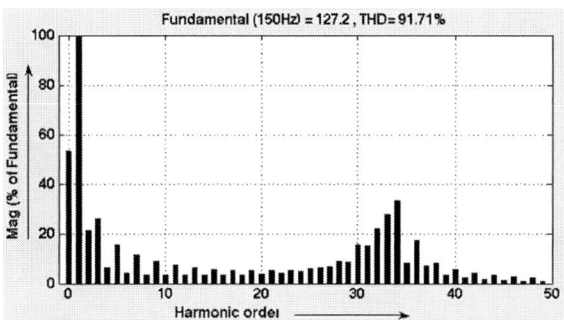

Figure 15 FFT analysis for output voltage with RL load

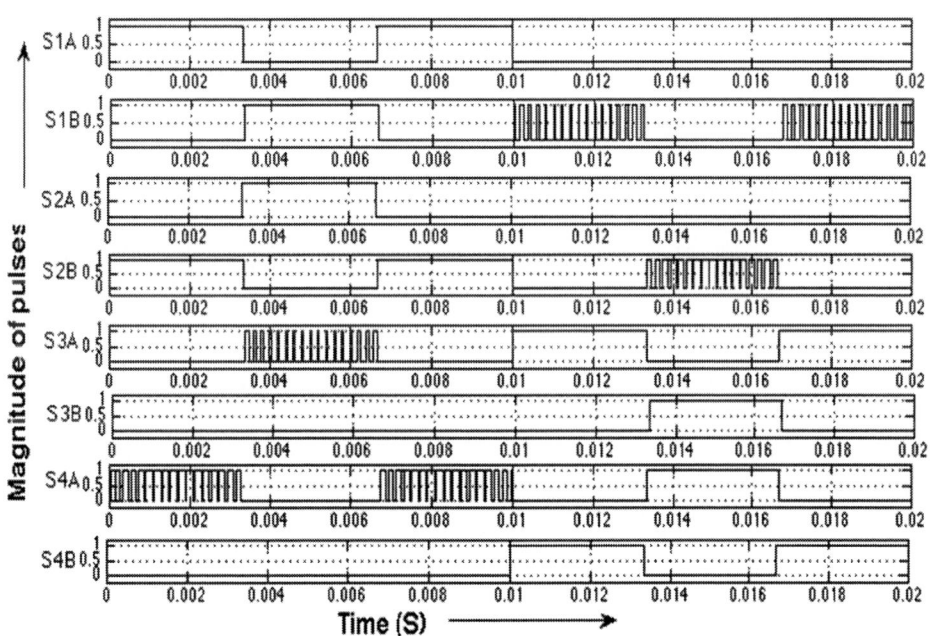

Figure 16 Switching signals for strategy II

Seventh International Conference on Advances in Power Electronics and Instrumentation Engineering – PEIE 2016

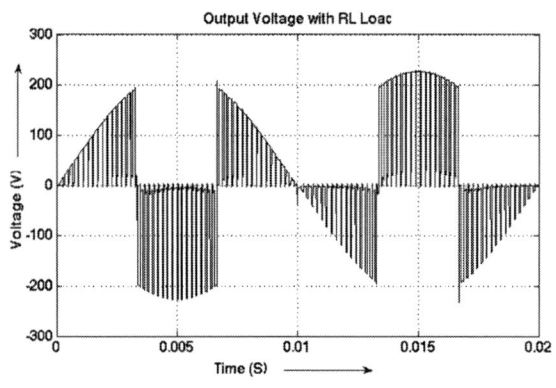

Figure 17 Output voltage for strategy II (RL load)

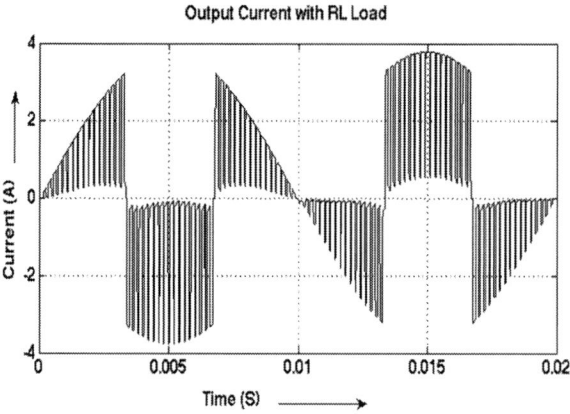

Figure 18 Output current for strategy II (RL load)

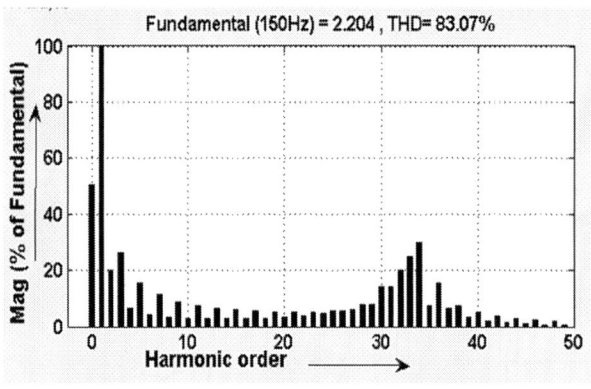

Figure 19 FFT analysis for output current with RL load

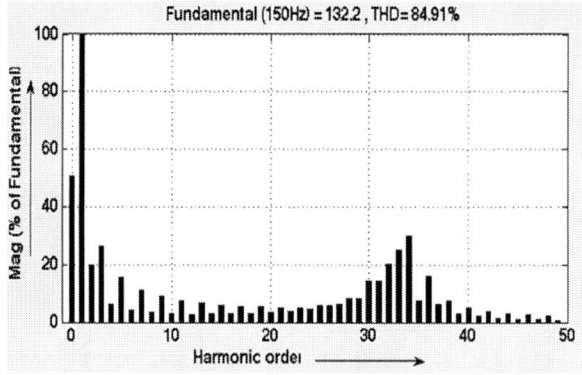

Figure 20 FFT analysis for output voltage with RL load

Strategy III - Proposed Strategy

In this strategy three switches will be used for a single time interval, out of which one switch will always be modulated with a continuous pulse and other two will be PWM switches. Thus one PWM signal is given to the switches used for free-wheeling (say PWM) and other PWM signal is given to the switch through which the load current would flow (say cPWM). These two PWM are generated such that one would be complementary of the other (only in that interval in which it is to be ON). Refer Figure 22, which clearly shows the three signals given to the switches of state 1. The two PWM signals in Figure 22 are zoomed in Figure 23, which shows that signal "cPWM" given to S2b is the complementary of signal "PWM" given to switch S4a, while S1a is modulated with a continuous pulse as shown in Figure 23. The switching sequence for proposed strategy is shown in Table 2.

- At any time 't', two switches S1a and S4a (PWM) will be kept ON and conduct the current flow during the positive cycle of input source, with S2b (cPWM) turn 'ON' for commutation purpose (state 1). [ref fig 24]
- At any time 't', two switches S4b and S1b (PWM) will be kept ON and conduct the current flow during the negative cycle of input source, with S3a (cPWM) turn 'ON' for commutation purpose (state 2). [ref fig 25]
- At any time 't', two switches S2a and S3a (PWM) will be kept ON and conduct the current flow during the positive cycle of input source, with S1b (cPWM) turn 'ON' for commutation purpose (state 3). [ref fig 26]
- At any time 't', two switches S3b and S2b (PWM) will be kept ON and conduct the current flow during the negative cycle of input source, with S4a (cPWM) turn 'ON' for commutation purpose (state 4). [ref fig 27]

As compared to strategy II we get an improved THD and also it can be used with both R load and RL load. Refer Figure 21 for the switching signals according to Table 2. The output voltage, current waveform and THD are shown in Figure 28 to Figure 31.

A Comparative Study of Switching Strategies for Single Phase Matrix Converter

Table 2: Switching sequence for strategy II and proposed strategy

I/P Frequency	Target O/P Frequency	Time Interval	State	For Strategy II		PWM Switch	For Strategy III (Proposed Strategy)		
				Switch modulated with continuous pulse			Switch Modulated with continues pulse	PWM Switch	
					For Free Wheeling			PWM	cPWM
50 Hz	150 Hz	1	1	S1a	S2b	S4a	S1a	S4a	S2b
		2	3	S2a	S1b	S3a	S2a	S3a	S1b
		3	1	S1a	S2b	S4a	S1a	S4a	S2b
		4	2	S4b	S3a	S1b	S4b	S1b	S3a
		5	4	S3b	S4a	S2b	S3b	S2b	S4a
		6	2	S4b	S3a	S1b	S4b	S1b	S3a

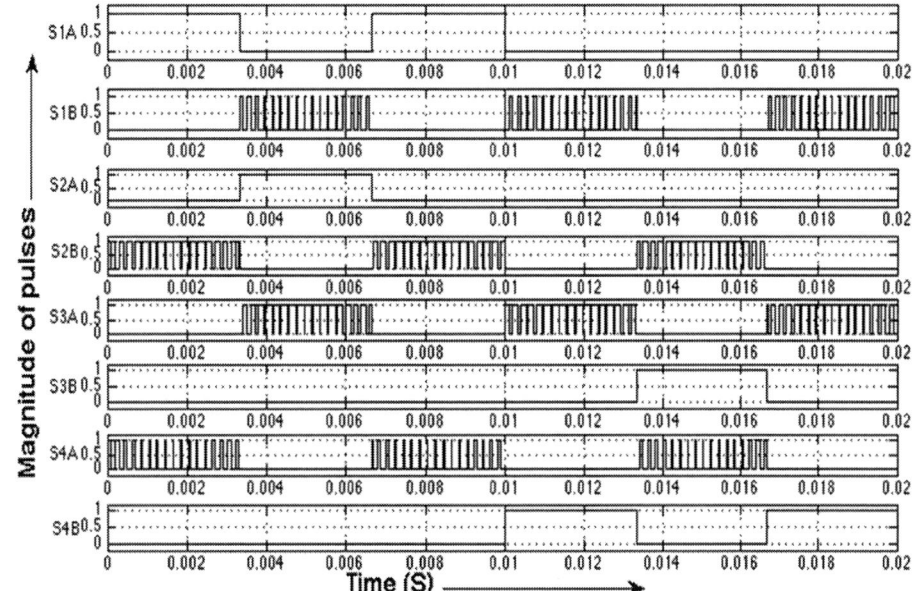

Figure 21: Switching signals for proposed strategy

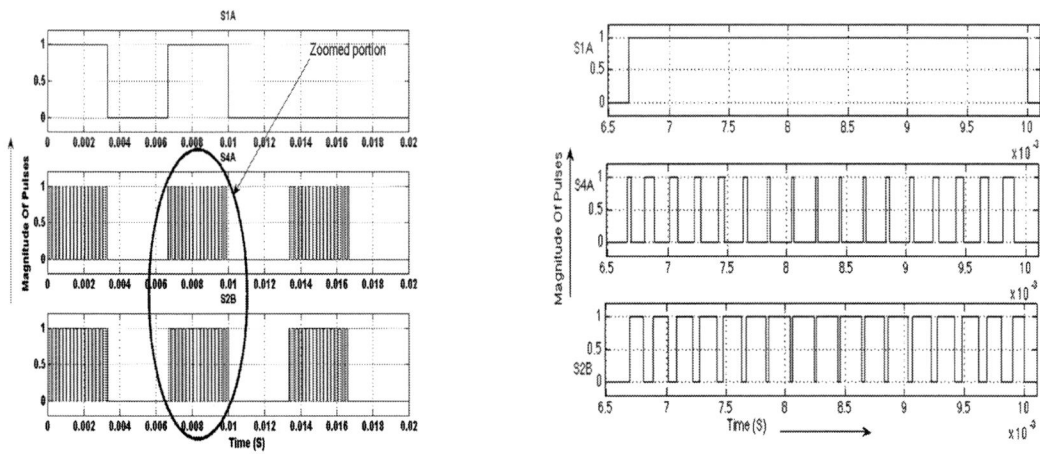

Figure 22 Switching signals for state 1 of proposed strategy   Figure 23 Zoomed view of Fig 22

Seventh International Conference on Advances in Power Electronics and Instrumentation Engineering – PEIE 2016

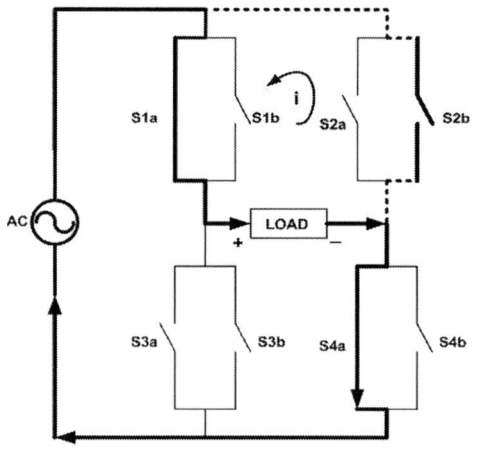

Figure 24 State 1 positive cycle

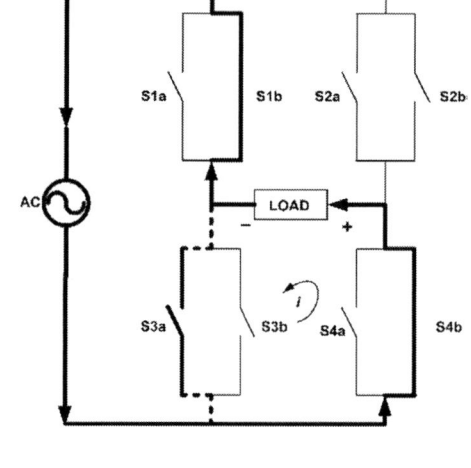

Figure 25 State 2 negative cycle

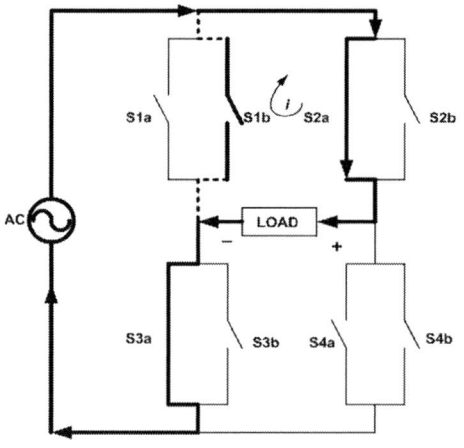

Figure 26 State 3 positive cycle

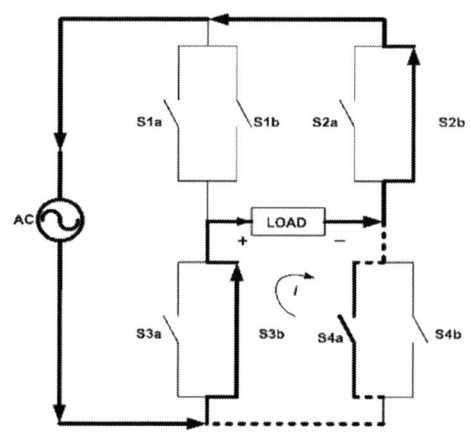

Figure 27 State 4 negative cycle

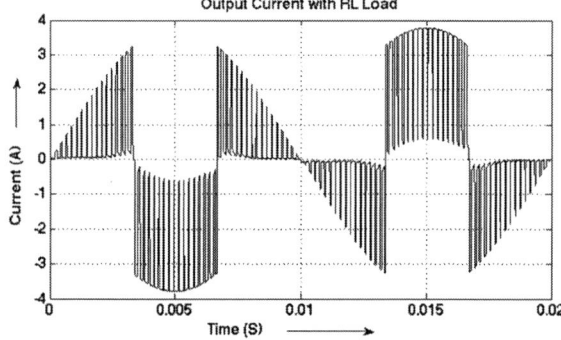

Figure 28 Output current for proposed strategy (RL load)

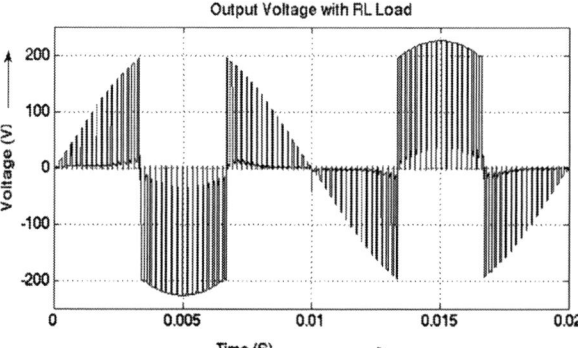

Figure 29 Output voltage proposed strategy (RL load)

A Comparative Study of Switching Strategies for Single Phase Matrix Converter

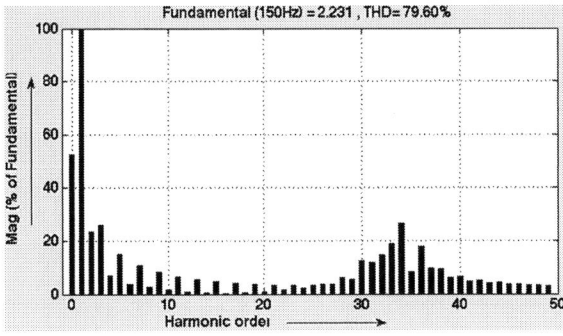

Figure 30 FFT analysis for output current

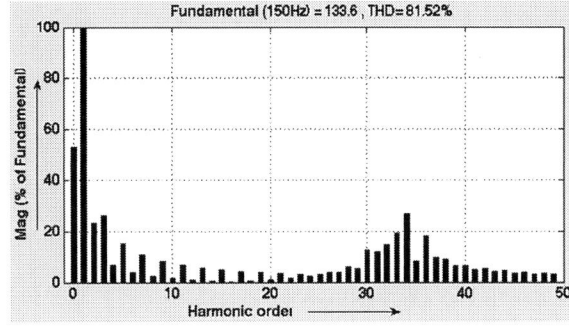

Figure 31 FFT analysis for output voltage

MATLAB Implementation

The matrix converter topology is developed in MATLAB/Simulink (MLS) software package as shown in Figure 32.

Table 3: THD results

Strategy	Current THD(%)	Voltage THD(%)
I (R Load)	91.59	91.59
I (RL Load)	85.05	91.71
II (RL Load)	83.08	84.91
III(RL Load) [Proposed Strategy]	79.60	81.52

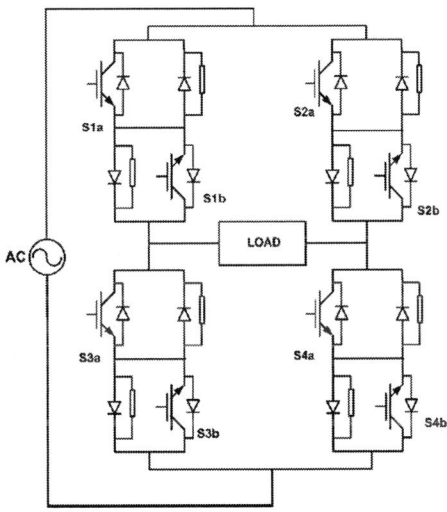

Figure 32 Simulation circuit diagram

The results in Table 3 shows that THD is improved with the proposed strategy. Moreover, it also provides us with a path for free-wheeling as three switches are kept ON for a single time interval.

Conclusion

Three different switching strategies have been presented in this paper. Comparison between all three strategies have been done and their advantages and limitations have been studied. The operational behavior is verified using MATLAB/Simulink with the SimPowerSystem Block Set. Results conclude that the strategy I can be used only for R load, strategy II can be used with R and RL load, proposed strategy (strategy III) provides better THD compared to strategy II.

Seventh International Conference on Advances in Power Electronics and Instrumentation Engineering – PEIE 2016

References

[1] Anand kumar, Dr. P.R. Thakura, "Hardware Development and Implementation of Single Phase Matrix Converter as a Cycloconverter and as an Inverter," Power Electronics (IICPE), IEEE 6th India International Conference, 2014.

[2] V.V.Subrahmanya Kumar Bhajana, Pavel Drabek, Martin Jara, Bedrich Bednar, "A Novel ZCS Single Phase Matrix converter for Traction Applications," Power Electronics and Applications (EPE'14-ECCE Europe), 16th European Conference, 2014.

[3] Zahirrudin Idris, Mustafar Kamal Hamzah, Ngah Ramzi Hamzah, "Modelling & Simulation of a new Single-phase to Single-phase Cycloconverter based on Single-phase Matrix Converter Topology with Sinusoidal Pulse Width Modulation Using MATLAB/Simulink", University Teknologi MARA, 40450 Shah Alam, Malaysia.

[4] Venturini, M., and Alesina, A., "The Generalized transformer: a new bi-directional sinusoidal wave form frequency converter with continuosly adjustable input power factor," IEEE Power Electron. Spec. Conf. Rec., 1980, pp. 242-252.

[5] Zuckerberger, A., Weinstock, D., Alexandrovitz A., "Single-phase Matrix Converter," IEE Proc. Electric Power App, Vol.144 (4), Jul 1997 pp. 235-240.

[6] Hosseini, S.H.; Babaei, E, "A new generalized direct matrix converter," IEEE International Symposium of Industrial Electronics, 2001. Proc. ISIE 2001. Vol (2), pp.1071-1076.

[7] Abdollah Koei & Subbaraya Yuvarajan, "Single-Phase AC-AC Converter Using Power Mosfet's," IEEE Transaction on Industrial Electronics, Vol. 35, No.3, August 1988, pp.442-443.

[8] Firdaus, S., Hamzah, M.K., "Modelling and simulation of a single-phase AC-AC matrix converter using SPWM," Proceedings on Student Conference on Research and Development, (SCOReD2002), 16-17 July 2002, pp.286-289.

[9] Empringham, L., Wheeler, P.W., Clare, J.C., "Intelligent Commutation of Matrix Converter Bi-directional Switch Cells Using Novel Gate Drive Techniques," 29th Annual IEEE Power Electronics Specialists Conference, 1998. PESC 98 Record., Vol.1, 17-22 May 1998, pp.707- 713.

[10] Kwon, B.-H., Min, B.-D., Kim, J.-H., "Novel Commutation Technique of AC-AC Converters," Electric Power Application, IEE Proceedings-, Vol. 145 (4), July 1998, pp. 295-300.

[11] Zahiruddin Idris, Siti Zaliha Mohammad Noor & Mustafar Kamal Hamzah, "Safe Commutation Strategy in Single-phase Matrix Converter", IEEE Sixth International Conference PEDS 2005, Kuala Lumpur, Malaysia.

[12] Karamat, A., Thomson, T., Mehta, P., "A novel strategy for control of cycloconverters," 22nd Annual EEEE Power Electronics Specialists Conference, 1991. PESC '91, Record. 24-27 June 1991, pp. 819 - 824.

[13] Choudhury, M.A., Uddin, M.B., Bhuyia, A.R., Rahman, M.A., "New topology and analysis of a single phase delta modulated cycloconverter," Proceedings of the 1996 International Conference on Power Electronics, Drives and Energy Systems for Industrial Growth, 8-11 Jan. 1996, Vol. 2, pp.819-825

[14] Hill, W.A., Ho, E.Y.Y., Nuezil, LJ. "Dynamic behavior of cyclo-converter system," IEEE Transactions on Industry Applications, Vol. 27(4), July-Aug. 1991, pp.750-755

[15] K.V.S Bharath, Ankit Bhardwaj, "Implementing Single Phase Cyclo-Converter Using Single Phase Matrix Converter Topology with Sinusoidal Pulse Width Modulation," International Journal for Technological Research in Engineering Volume 2, Issue 6, February-2015.

[16] Wheeler, P.W., Rodriguez, J., Clare, J.C., Empringham, L., Weinstein, A., "Matrix converters: a technology review," Industrial Electronics, IEEE Transactions on, Volume: 49, Issue: 2, April 2002 Pages: 276 – 288

[17] Wheeler, P.W., Clare, J.C., Empringham, L., Bland, M., Kerris, K.G., "Matrix converters," Industry Applications Magazine, IEEE, Volume: 10, Issue: 1, Jan-Feb2004 Pages: 59 – 65.

[18] Cho, J.G., and Cho, G.H, "Soft-switched Matrix Converter for High Frequency direct AC-to-AC Power Conversion," Int. J. Electron., 1992, 72, (4), pp. 669-680.

[19] Idris, Z.; Hamzah, M.K.; Saidon, M.F., "Implementation of Single-Phase Matrix Converter as a Direct AC-AC Converter with Commutation Strategies," 37th IEEE Power Electronics Specialists Conference, 2006. PESC '06, Page(s):1-7.

[20] H.M.Hanafi, N.R. Hamzah, A.Saparon and M.K.Hamzah, "Improved Switching Strategy of Single-Phase Matrix Converter as a Direct AC-AC Converter", University Teknologi MARA, 40450 Shah Alam, Malaysia.

Author Index

H

Hormaz Amrolia   18, 29

K

Kapil P N   18
Krishna R More   18

M

Mohammadamin Yusufji Khatri   29

P

Pulkit Singh   1
Palwalia D K   1

Grenze Scientific Society
Jyothi Nagar-48/1
Kesavadasapuram, Trivandrum - 695004

ISBN 978-1-5108-6066-7